Credit Risk

Modelling credit risk accurately is central to the practice of mathematical finance. The majority of available texts are aimed at an advanced level, and are more suitable for PhD students and researchers. This volume of the Mastering Mathematical Finance series addresses the need for a course intended for master's students, final-year undergraduates, and practitioners. The book focuses on the two mainstream modelling approaches to credit risk, namely structural models and reduced-form models, and on pricing selected credit risk derivatives. Balancing rigorous theory with examples, it takes readers through a natural development of mathematical ideas and financial intuition.

MAREK CAPIŃSKI is Professor of Applied Mathematics at AGH University of Science and Technology, Kraków. His research interests include mathematical finance, corporate finance, and hydrodynamics. He has been teaching for over 35 years, has held visiting fellowships in Poland and the UK, and has published over fifty research papers and nine books.

TOMASZ ZASTAWNIAK is Chair in Mathematical Finance at the University of York. His research interests include mathematical finance, stochastic analysis, stochastic optimisation and convex analysis, and mathematical physics. He has previously taught at numerous institutions in Poland, the USA, Canada, and the UK, and has published over fifty research publications and eight books.

T0353506

Mastering Mathematical Finance

Mastering Mathematical Finance is a series of short books that cover all core topics and the most common electives offered in Master's programmes in mathematical or quantitative finance. The books are closely coordinated and largely self-contained, and can be used efficiently in combination but also individually.

The MMF books start financially from scratch and mathematically assume only undergraduate calculus, linear algebra, and elementary probability theory. The necessary mathematics is developed rigorously, with emphasis on a natural development of mathematical ideas and financial intuition, and the readers quickly see real-life financial applications, both for motivation and as the ultimate end for the theory. All books are written for both teaching and self-study, with worked examples, exercises, and solutions.

[DMFM] *Discrete Models of Financial Markets*,
        Marek Capiński, Ekkehard Kopp

[PF]   *Probability for Finance*,
        Ekkehard Kopp, Jan Malczak, Tomasz Zastawniak

[SCF]   *Stochastic Calculus for Finance*,
        Marek Capiński, Ekkehard Kopp, Janusz Traple

[BSM]   *The Black–Scholes Model*,
        Marek Capiński, Ekkehard Kopp

[PTRM] *Portfolio Theory and Risk Management*,
        Maciej J. Capiński, Ekkehard Kopp

[NMFC] *Numerical Methods in Finance with C++*,
        Maciej J. Capiński, Tomasz Zastawniak

[SIR]   *Stochastic Interest Rates*,
        Daragh McInerney, Tomasz Zastawniak

[CR]   *Credit Risk*,
        Marek Capiński, Tomasz Zastawniak

[SCAF] *Stochastic Control Applied to Finance*,
        Szymon Peszat, Tomasz Zastawniak

Series editors Marek Capiński, *AGH University of Science and Technology, Kraków*; Ekkehard Kopp, *University of Hull*; Tomasz Zastawniak, *University of York*

Credit Risk

MAREK CAPIŃSKI
AGH University of Science and Technology, Kraków

TOMASZ ZASTAWNIAK
University of York

CAMBRIDGE
UNIVERSITY PRESS

University Printing House, Cambridge CB2 8BS, United Kingdom

One Liberty Plaza, 20th Floor, New York, NY 10006, USA

477 Williamstown Road, Port Melbourne, VIC 3207, Australia

4843/24, 2nd Floor, Ansari Road, Daryaganj, Delhi - 110002, India

79 Anson Road, #06-04/06, Singapore 079906

Cambridge University Press is part of the University of Cambridge.

It furthers the University's mission by disseminating knowledge in the pursuit of
education, learning and research at the highest international levels of excellence.

www.cambridge.org
Information on this title: www.cambridge.org/9781107002760
10.1017/9781139047432

First published 2017

A catalogue record for this publication is available from the British Library

ISBN 978-1-107-00276-0 Hardback
ISBN 978-0-521-17575-3 Paperback

Additional resources for this publication at www.cambridge.org/creditrisk.

Contents

Preface

Credit risk is concerned with the possibility of bankruptcy happening as a result of unpaid debt obligations. This applies to companies partially financed by debt. The key question is to find the value of debt or, equivalently, the amount of interest charged by the debt holders to offset their possible loss due to bankruptcy. Our goal is to explain the main issues in the simplest possible cases, to keep the theory free of technical complications.

The first approach covered in this book relates bankruptcy to the company's performance. The resulting model is referred to as structural. Historically, this was the first systematic approach to credit risk, developed by Merton in 1974.

In the other method studied here, called the reduced-form approach, default occurs as a result of external forces, not related directly to the functioning of the company, and comes as a surprise at a random time. This may be the time when some information is revealed about technological advances, court decisions, or political events, which could wipe out the company.

We develop an approach similar to the classical financial theories, where certain underlying securities are specified and self-financing strategies are constructed from these. Here it is the risk-free and defaultable zero-coupon bonds that will serve as underlying assets. This makes it possible to price a range of defaultable securities, including Credit Default Swaps, by means of replication. Then, after adding a Black–Scholes stock as a building block of the market, the range of securities broadens, and this allows us to conclude with an example where the structural and reduced-form approaches come together.

Readers of the book need to be familiar with material covered in some earlier volumes in this series, namely [SCF] and [BSM], that is, with the foundations of stochastic calculus and the Black–Scholes model. Exposure to measure theory based probability, covered in [PF], would also be helpful.

The text is interspersed with various examples and exercises. The numerical work underpinning many of the examples, and solutions to the exercises, are available at www.cambridge.org/creditrisk.

1

Structural models

Consider a company launched at time 0, when some assets are purchased for $V(0)$. Funding comes from two sources. Shareholders contribute $E(0)$, referred to as **equity**. The remaining amount $D(0) = V(0) - E(0)$, called **debt**, is either borrowed from a bank or raised by selling bonds issued by the company.

We consider this company over a time interval from 0 to T, during which the assets are put to work in order to generate some funds, which are then split between the two groups of investors at time T. The debt is first repaid with interest to the debt holders, who have priority over the equity holders. Any remaining amount goes to the equity holders.

The simplest way to raise money to make these payments is to sell the assets of the company. We begin our analysis with this case, by making the necessary assumption that the assets are tradeable.

1.1 Traded assets

We assume that there is a liquid market for the assets, and $V(t)$ for $t \in [0, T]$ represents their market value. We also assume that the assets generate no additional cash flows. A practical example of such assets would be a portfolio of traded stocks that pay no dividends, the company being an investment fund.

Payoffs

Suppose that the company has to clear the debt at time T, and that there are no intermediate cash flows to the debt holders. The interest rate applying to the loan will be quoted by the bank or implied by the bond price. We denote this **loan rate** by k_D with continuous compounding, and by K_D with annual compounding. The amount due at time T is

$$F = D(0)e^{k_D T} = D(0)(1 + K_D)^T.$$

(Throughout this volume we take one year as the unit of time.) One of the goals here is to find the loan rate that reflects the risk for the debt holders.

At time T we sell the assets and close down the business, at least hypothetically, to analyse the company's financial position at that time. It is possible that the amount obtained by selling the assets is insufficient to settle the debt, that is, $V(T) < F$. In this respect, we make an important assumption concerning the legal status of the company: it has **limited liability**. This means that losses cannot exceed the initial equity value $E(0)$. If $V(T) < F$, the equity holders do not have to cover the loss from their personal funds. The company is declared bankrupt, and the equity holders walk away having lost their initial investment. If $V(T) \geq F$, the loan can be paid back with interest, and the equity holders keep the balance. The final value of equity is therefore a random amount equal to the payoff of a call option,

$$E(T) = \max\{V(T) - F, 0\},$$

with the value of the assets as the underlying security and the debt repayment amount F as the strike price.

Remark 1.1
A call is an option to buy the underlying asset for a prescribed price, which sounds paradoxical here. However, it is consistent with the general practice that loans are secured on some assets. The borrower's ownership rights in the assets are restricted until the loan is repaid. The full rights (for instance, to sell the asset) are in a sense bought back when the loan is settled.

If $V(T) \geq F$, the debt will be paid in full, but in the case of bankruptcy, which will be declared if $V(T) < F$, the debt holders are going to take over the assets and sell them for $V(T)$. This is straightforward if the assets are tradeable. The amount received at time T will be

$$D(T) = \min\{F, V(T)\} = F - \max\{F - V(T), 0\}.$$

We immediately recognise the put payoff as one of the components,

$$P(T) = \max\{F - V(T), 0\}.$$

It reflects the limited liability feature.

Observe that

$$D(T) + E(T) = \min\{F, V(T)\} + \max\{V(T) - F, 0\} = V(T).$$

This is similar to the equality $V(0) = D(0) + E(0)$, which holds at time 0, and is called the **balance sheet equation**. It illustrates one of the basic rules of corporate finance: the assets are equal to the liabilities (debt plus equity).

If the payoff of the put option is identically zero (which is possible, for example in the binomial model when the strike price is low enough), then $D(T) = F$ and the debt position is risk free. In this case, we should have $k_D = r$, the continuously compounded risk-free rate. If it is possible that the debt holders recover less than F, a higher rate k_D will be applied to compensate for the risk.

These remarks motivate the following proposition, which does not depend on any particular model for the asset value process. Let $P(0)$ denote the time 0 price of the put option.

Proposition 1.2
If $P(0) > 0$, then $k_D > r$.

Proof Recall the put-call parity relationship expressed in terms of a call, put, stock, and a general strike price K:

$$S(0) = C(0) - P(0) + Ke^{-rT}.$$

In our case it becomes

$$V(0) = E(0) - P(0) + Fe^{-rT}.$$

This implies that

$$D(0) = Fe^{-rT} - P(0).$$

In other words,

$$F = [D(0) + P(0)]e^{rT}. \tag{1.1}$$

Recall that $F = D(0)e^{k_D T}$, so

$$D(0)e^{k_D T} = [D(0) + P(0)]e^{rT},$$

which yields

$$k_D = r + \frac{1}{T} \ln\left(1 + \frac{P(0)}{D(0)}\right)$$

and implies that $k_D > r$ as claimed since the second term on the right-hand side is positive.   □

As we can see, the loan rate k_D is typically higher than the risk-free rate r. This is consistent with intuition since the loan is not free of risk as the full amount F is paid only in some circumstances.

Definition 1.3

The difference $s = k_D - r$ is called the **credit spread**.

This quantity represents the additional return demanded by the debt holders to compensate for their exposure to default risk.

The positivity of the credit spread is all that can be discovered without specifying a model for asset values. Such a model is needed to have a method of computing option prices, and in turn solving the pivotal problem of setting the level of F, hence k_D, for a given debt and equity values $E(0)$ and $D(0)$, which determine the financial structure of the company.

It is often convenient to describe the financial structure of the company in terms of ratios rather than the actual debt and equity values.

Definition 1.4

The **debt** and **equity ratios** are defined as

$$w_D = \frac{D(0)}{V(0)}, \quad w_E = \frac{E(0)}{V(0)}.$$

Because $V(0) = D(0) + E(0)$, these ratios satisfy $w_E + w_D = 1$.

We have seen that equity can be regarded as a call option with strike F. The time 0 price of this call option, which we now denote by $C(F)$, satisfies $C(F) = E(0)$. This equation can be solved for F, and we illustrate this with the simplest model.

Binomial model

Consider the single-step binomial model and suppose that $V(T)$ takes only two values $V(0)(1 + U)$ and $V(0)(1 + D)$ determined by the returns $-1 < D < R < U$, where R is the risk-free return, that is, $1 + R = e^{rT}$. The non-trivial range of strike prices is

$$V(0)(1 + D) < F < V(0)(1 + U),$$

and then

$$E(0) = C(F) = \frac{1}{1+R}q(V(0)(1 + U) - F),$$

where

$$q = \frac{R - D}{U - D}$$

is the risk-neutral probability (see [DMFM]). The formula is justified by the fact that the payoff of the call option can be replicated by means of $V(T)$ and the risk-free asset, both assumed tradeable. This gives us the corresponding range of initial equity values

$$0 < E(0) < \frac{1}{1+R}qV(0)(U - D).$$

The equation for F can be solved to get

$$F = V(0)(1 + U) - \frac{1}{q}E(0)(1 + R),$$

and then

$$e^{k_D T} = (1 + K_D)^T = \frac{F}{D(0)}.$$

The investors are interested in real-life probabilities to evaluate their prospects, and these should be used to find the expected returns and standard deviations of returns for equity and debt. The computations are straightforward, and we simply consider a numerical example.

Example 1.5

Let $V(0) = 100$, $T = 1$, $R = 20\%$, $U = 40\%$, and $D = -40\%$, hence $q = 0.75$. With $E(0) = 40$ we find $F = 76$ and $K_D = 26.67\%$. Assuming the real-life probability of the up movement to be $p = 0.9$, we get the expected return on assets to be $\mu_V = 32\%$, the expected return on equity $\mu_E = 44\%$, and on debt $\mu_D = 24\%$. The last figure is important for the debt holders as it will be earned on average if the loan rate K_D is quoted for all similar customers. The standard deviation of the return on debt is $\sigma_D = 8\%$.

If equity is reduced to $E(0) = 20$, we have $F = 108$, $K_D = 35\%$, and the expected return on debt under the real-life probability grows to 29% while the standard deviation of the return on debt becomes 18%.

In the extreme (and unrealistic) case of $E(0) = 50$ we have $K_D = R$ and the expected return is the same, the risk being zero (the debt payoff is F in each scenario). In the other extreme case of $E(0) = 0$ the company is owned entirely by the debt holders, the parameters for debt coinciding with those for the assets.

Since the payoff for equity is an affine function of $V(T)$, the expected return on equity is the same for each level of financing, $\mu_E = 44\%$. The model is not sophisticated enough to see anything interesting here.

Remark 1.6

In the single-step binomial model the balance between the expected return μ_H and standard deviation σ_H of return of any derivative security H (in particular $H = V, E$, or D) is captured by the fact that the market price of risk $\frac{\mu_H - R}{\sigma_H}$ is the same for each H; see [DMFM].

For two steps the situation becomes more interesting. There are three possible values of $V(T)$ and larger scope for non-trivial cases. The range for the strike price is

$$V(0)(1 + D)^2 < F < V(0)(1 + U)^2.$$

The pricing formula and, in particular, the equation for F become more complicated. Once again, we just analyse a numerical example.

Example 1.7

Using the data from Example 1.5, we perform computations for two cases, $E(0) = 40$ and $E(0) = 60$. We find the respective values of F to be 93.60 and 59.04, the expected two-period returns on equity 107.36% and 92.38%, and the corresponding standard deviations 100.43% and 74.20%. The corresponding expected returns on debt are 52.16% and 47.02% (as compared with the risk-free return of 44%), with standard deviations 11.11% and 5.73%, respectively.

Exercise 1.1 Derive an explicit general formula for F in the two-step binomial model, and compute the expected return and standard deviation of the return for equity and debt as in Example 1.7 in the case of 50% financing by equity.

Rather than considering the n-step binomial model and using the Cox–Ross–Rubinstein (CRR) formula for the call price (see [DMFM]), which is similar to the Black–Scholes formula, we proceed directly to the latter as it is important and in fact easier to handle.

1.2 Merton model

Suppose that the assets of a company are tradeable and follow the Black–Scholes model, i.e. satisfy the stochastic differential equation

$$dV(t) = \mu V(t)dt + \sigma V(t)dW_P(t),$$

where $W_P(t)$ is a Wiener process under the real-life probability P. Girsanov's theorem (see [BSM]) makes it possible to change to the risk-neutral probability Q and write the stochastic differential equation as

$$dV(t) = rV(t)dt + \sigma V(t)dW_Q(t),$$

where W_Q is a Wiener process under Q.

Let us put

$$C(V(0), \sigma, r, T, F) = V(0)N(d_+) - e^{-rT}FN(d_-),$$

where

$$d_+ = \frac{\ln \frac{V(0)}{F} + (r + \frac{1}{2}\sigma^2)T}{\sigma \sqrt{T}}, \quad d_- = \frac{\ln \frac{V(0)}{F} + (r - \frac{1}{2}\sigma^2)T}{\sigma \sqrt{T}}, \quad (1.2)$$

and where

$$N(x) = \int_{-\infty}^{x} \frac{1}{\sqrt{2\pi}} e^{\frac{1}{2}y^2} dy$$

is the standard normal cumulative distribution function. We recognise the expression defining $C(V(0), \sigma, r, T, F)$ as the Black–Scholes call pricing formula; see [BSM].

We have $E(0) = C(V(0), \sigma, r, T, F)$ since equity is a call option with strike F. This can be written as

$$E(0) = V(0)N(d_+) - e^{-rT}FN(d_-).$$

The equation needs to be solved for F numerically, with $V(0)$, σ, r, and T fixed. (When solving the equation, remember that d_+ and d_- also depend on F.) The formula for the initial value of debt reads

$$D(0) = V(0)N(-d_+) + e^{-rT}FN(d_-).$$

This follows from the balance sheet equation $V(0) = E(0) + D(0)$ and the symmetry $1 - N(x) = N(-x)$ of the standard normal distribution. Together, these are the ingredients of **Merton's model** of credit risk.

Example 1.8
Let $V(0) = 100$ and consider 50% financing by equity. Assume the risk-free rate $r = 5\%$ and volatility $\sigma = 30\%$, and take $T = 1$. We can solve the equation

$$C(V(0), \sigma, r, T, F) = 50$$

to find $F = 52.6432$, and then compute the loan rate

$$k_D = \frac{1}{T} \ln \frac{F}{D(0)} = 5.1515\%,$$
$$K_D = e^{k_D} - 1 = 5.2865\%.$$

Exercise 1.2   Within the setup of Example 1.8 consider an investment in stock with volatility higher than 30%. What does your intuition say about the impact of this on k_D? Analyse the monotonicity of k_D as a function of σ. Perform numerical computations for $\sigma = 35\%$.

Expected returns

It is interesting to find the expected returns on equity and debt between the time instants 0 and T under the real-life probability P and analyse their dependence on the financial structure. To this end we need to compute the expectation $\mathbb{E}_P(E(T))$ under the real-life probability. The Black–Scholes formula gives a similar expectation but under the risk-neutral probability,

$$\mathbb{E}_Q(E(T)) = \mathbb{E}_Q((V(T) - F)^+) = e^{rT} C(V(0), \sigma, r, T, F),$$

where

$$V(T) = V(0) \exp\left(\left(r - \frac{1}{2}\sigma^2\right)T + \sigma W_Q(T)\right).$$

This formula is valid for every $r > 0$, in particular for $r = \mu$. On the other hand, we also have

$$V(T) = V(0) \exp\left(\left(\mu - \frac{1}{2}\sigma^2\right)T + \sigma W_P(T)\right).$$

Because $W_Q(T)$ has the same probability distribution under the risk-neutral probability Q as $W_P(T)$ under the real-life probability P (namely the normal distribution $N(0, T)$), it follows that

$$\mathbb{E}_P(E(T)) = \mathbb{E}_P((V(T) - F)^+) = e^{\mu T} C(V(0), \sigma, \mu, T, F).$$

This enables us to compute the expected return on equity under the real-life probability,

$$\mu_E = \frac{\mathbb{E}_P(E(T)) - E(0)}{E(0)}.$$

To find the expected return on debt μ_D we can use the relationship

$$\mu_V = w_E \mu_E + w_D \mu_D$$

from portfolio theory (see [PTRM]), with $\mu_V = e^{\mu T} - 1$ since $\mathbb{E}_P(V(T)) = V(0)e^{\mu T}$.

Example 1.9

For the data from Example 1.8 and $\mu = 10\%$ we get $\mu_V = 10.52\%$, $\mu_E = 15.85\%$, and $\mu_D = 5.19\%$.

Exercise 1.3 Using the data in Example 1.8 and $\mu = 10\%$, compute F and μ_E, μ_D for a company with 40% financing by equity, and also for one with 60% financing by equity.

Example 1.10

For the data in Example 1.8 and $\mu = 10\%$, in Figure 1.1 we plot the graphs of the expected returns μ_E and μ_D as functions of the equity ratio w_E. For comparison, we include the expected return $\mu_V = e^{\mu T} - 1$ on the assets, independent of financing, thus a horizontal line.

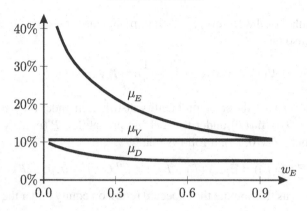

Figure 1.1 Expected returns μ_V, μ_E, μ_D as functions of equity ratio w_E.

High level of debt is profitable for equity holders, and it also appears attractive to debt holders. However, in real life the amount of debt recovered following bankruptcy will be reduced by legal costs. In addition, rapid liquidation of a large number of assets may reduce the prices. These factors affect the debt payoff and hence the expected return on debt μ_D, computed above assuming full recovery.

Partial recovery

We are going to discuss the case when the market value of the company's assets cannot be fully recovered due to the cost of bankruptcy procedures. It is not possible to use the relationship $\mu_V = w_D\mu_D + w_E\mu_E$ from portfolio theory because additional participants emerge in the case of bankruptcy, such as bailiffs or legal services providers.

Suppose that the amount recovered by debt holders is proportional to the value of the assets. When default occurs, that is, when $V(T) < F$, bankruptcy procedures are initiated, the assets are sold, and the debt holders receive $\alpha V(T)$, where $\alpha \in [0, 1]$ is a constant **recovery rate**. Otherwise, when $V(T) \geq F$, the company remains solvent and able to settle the debt in full. As a result, the debt payoff becomes

$$D(T) = F\mathbf{1}_{\{V(T)\geq F\}} + \alpha V(T)\mathbf{1}_{\{V(T)<F\}}.$$

When $\alpha < 1$ the debt holders' payoff $\alpha V(T)$ in the case of bankruptcy is reduced as compared to the full recovery payoff $V(T)$ in the case with $\alpha = 1$. To be compensated for this reduction, the debt holders will demand a higher loan rate k_D, that is, a higher amount $F = D(0)e^{k_D T}$ due at maturity if there is no default (the sum borrowed, $D(0)$, remains unchanged). This amount F can be found by expressing the time 0 value of debt as the discounted expectation of the debt payoff $D(T)$ under the risk-neutral probability Q, which we denote by

$$D(V(0), \sigma, r, T, F) = e^{-rT} \mathbb{E}_Q(D(T)),$$

and by solving the equation

$$D(0) = D(V(0), \sigma, r, T, F) \tag{1.3}$$

for F. For better clarity we denote the solution by F_α, so the above equation becomes

$$D(0) = e^{-rT} \mathbb{E}_Q(F_\alpha \mathbf{1}_{\{V(T) \geq F_\alpha\}} + \alpha V(T) \mathbf{1}_{\{V(T) < F_\alpha\}}).$$

The formula for equity payoff is, as before, that of a call option, with modified strike F_α. Namely,

$$E(T) = \max \{V(T) - F_\alpha, 0\}.$$

The value of this payoff is reduced due to the fact that the strike will be set higher in the case of partial recovery as compared to full recovery. Initially, the company is financed by an amount $E(0)$ from equity holders and by debt $D(0)$. As soon as the company is set up at time 0 by entering into a debt agreement and investing the total $V(0) = E(0) + D(0)$ into the assets, the value of equity drops down to a new level

$$E_\alpha(0) = e^{-rT} \mathbb{E}_Q(E(T)) = C(V(0), \sigma, r, T, F_\alpha),$$

which is lower than the invested amount $E(0)$. Indeed, the difference is

$$
\begin{aligned}
E(0) - E_\alpha(0) &= V(0) - D(0) - e^{-rT} \mathbb{E}_Q(\max \{V(T) - F_\alpha, 0\}) \\
&= e^{-rT} \mathbb{E}_Q(V(T) - D(T) - \max \{V(T) - F_\alpha, 0\}) \\
&= e^{-rT} \mathbb{E}_Q(V(T) - F_\alpha \mathbf{1}_{\{V(T) \geq F_\alpha\}} - \alpha V(T) \mathbf{1}_{\{V(T) < F_\alpha\}} \\
&\quad - (V(T) - F_\alpha) \mathbf{1}_{\{V(T) \geq F_\alpha\}}) \\
&= e^{-rT} \mathbb{E}_Q((1 - \alpha)V(T) \mathbf{1}_{\{V(T) < F_\alpha\}}) \geq 0.
\end{aligned}
$$

We recognise $(1 - \alpha)V(T) \mathbf{1}_{\{V(T) < F_\alpha\}}$ as the cost of liquidating the company in the case of a default. Note that $V(0) > E_\alpha(0) + D(0)$.

To compute F_α we need to derive a formula for $D(V(0), \sigma, r, T, F_\alpha)$ and then solve equation (1.3). It is not difficult to compute the expectation

$$\mathbb{E}_Q(D(T)) = F_\alpha Q(V(T) \geq F_\alpha) + \alpha \mathbb{E}_Q(V(T) \mathbf{1}_{\{V(T) < F_\alpha\}}).$$

Writing

$$V(T) = V(0)e^{(r-\frac{1}{2}\sigma^2)T+\sigma\sqrt{T}X},$$

where X follows the standard normal distribution $N(0, 1)$ under Q, we have

$$Q(V(T) \geq F_\alpha) = Q(-X \leq d_-),$$

$$\mathbb{E}_Q(V(T)\mathbf{1}_{\{V(T)<F_\alpha\}}) = V(0)\mathbb{E}_Q\left(e^{(r-\frac{1}{2}\sigma^2)T+\sigma\sqrt{T}X}\mathbf{1}_{\{X<-d_-\}}\right),$$

with d_- given by (1.2) with F_α replacing F. Next

$$Q(V(T) \geq F_\alpha) = N(d_-),$$

$$\mathbb{E}_Q(V(T)\mathbf{1}_{\{V(T)<F_\alpha\}}) = V(0)e^{(r-\frac{1}{2}\sigma^2)T}\int_{-\infty}^{-d_-}\frac{1}{\sqrt{2\pi}}e^{\frac{1}{2}x^2}e^{\sigma\sqrt{T}x}dx.$$

The integral can be evaluated in a similar way as in the derivation of the Black–Scholes formula. For the convenience of the reader we present the calculations:

$$\int_{-\infty}^{-d_-}\frac{1}{\sqrt{2\pi}}e^{\frac{1}{2}x^2}e^{\sigma\sqrt{T}x}dx = \frac{1}{\sqrt{2\pi}}e^{\frac{1}{2}\sigma^2 T}\int_{-\infty}^{-d_-}e^{-\frac{1}{2}(x-\sigma\sqrt{T})^2}dx$$

$$= \frac{1}{\sqrt{2\pi}}e^{\frac{1}{2}\sigma^2 T}\int_{-\infty}^{-d_+}e^{-\frac{1}{2}y^2}dy$$

$$= e^{\frac{1}{2}\sigma^2 T}N(-d_+),$$

where we use the substitution $y = x - \sigma\sqrt{T}$, and where d_+ is given by (1.2) with F_α replacing F. As a result,

$$D(V(0), \sigma, r, T, F_\alpha) = e^{-rT}F_\alpha N(d_-) + \alpha V(0)N(-d_+).$$

We are ready to compute F_α by solving (1.3) numerically.

Example 1.11
For the data in Example 1.10, with debt and equity financing at $w_D = w_E = 50\%$, we compute the final value F_α of debt at maturity and the corresponding value of equity $E_\alpha(0)$ at time 0 for various values of the recovery rate α ranging from 0.5 to 1.

α	0.5	0.6	0.7	0.8	0.9	1.0
F_α	53.0465	52.9611	52.8782	52.7977	52.7194	52.6432
$E_\alpha(0)$	49.6225	49.7025	49.7801	49.8554	49.9287	50.0000

Risk

We proceed to the task of computing the standard deviations of the returns on equity and debt to gain some additional insight.

We begin with computing the expected squared equity payoff

$$
\begin{aligned}
\mathbb{E}_P(E(T)^2) &= \mathbb{E}_P(((V(T) - F)^+)^2) \\
&= \mathbb{E}_P(\mathbf{1}_{\{V(T) \geq F\}}(V^2(T) - 2V(T)F + F^2)) \\
&= V(0)^2 e^{2(\mu - \frac{1}{2}\sigma^2)T} \int_d^\infty \frac{1}{\sqrt{2\pi}} e^{-\frac{1}{2}x^2} e^{2\sigma\sqrt{T}x} dx \\
&\quad - 2FV(0)e^{(\mu - \frac{1}{2}\sigma^2)T} \int_d^\infty \frac{1}{\sqrt{2\pi}} e^{-\frac{1}{2}x^2} e^{\sigma\sqrt{T}x} dx + F^2(1 - N(d)).
\end{aligned}
$$

Next, similar to before, we evaluate the integral

$$
\begin{aligned}
\int_d^\infty \frac{1}{\sqrt{2\pi}} e^{-\frac{1}{2}x^2} e^{2\sigma\sqrt{T}x} dx &= \frac{1}{\sqrt{2\pi}} \int_d^\infty e^{-\frac{1}{2}(x^2 - 4\sigma\sqrt{T}x + 4\sigma^2 T)} e^{2\sigma^2 T} dx \\
&= e^{2\sigma^2 T} \frac{1}{\sqrt{2\pi}} \int_d^\infty e^{-\frac{1}{2}(x - 2\sigma\sqrt{T})^2} dx \\
&= e^{2\sigma^2 T} \frac{1}{\sqrt{2\pi}} \int_{d - 2\sigma\sqrt{T}}^\infty e^{-\frac{1}{2}y^2} dy \\
&= e^{2\sigma^2 T}(1 - N(d - 2\sigma\sqrt{T})),
\end{aligned}
$$

where we make the substitution $y = x - 2\sigma\sqrt{T}$. Hence, replacing 2σ by σ, we get

$$
\int_d^\infty \frac{1}{\sqrt{2\pi}} e^{-\frac{1}{2}x^2} e^{\sigma\sqrt{T}x} dx = e^{\frac{1}{2}\sigma^2 T}(1 - N(d - \sigma\sqrt{T})).
$$

These formulae give the standard deviation of the return on equity

$$
\begin{aligned}
\sigma_E^2 &= \mathbb{E}_P\left(\left(\frac{E(T) - E(0)}{E(0)}\right)^2\right) - \mu_E^2 \\
&= \frac{\mathbb{E}_P(E(T)^2)}{E(0)^2} - \frac{2\mathbb{E}_P(E(T))}{E(0)} + 1 - \mu_E^2,
\end{aligned}
$$

where $\mathbb{E}_P(E(T)) = E(0)(1 + \mu_E)$ with μ_E computed as before.

For debt we work along similar lines and assume full recovery. First we find

$$\mathbb{E}_P(D(T)^2) = \mathbb{E}_P(\min\{V(T), F\}^2)$$
$$= \mathbb{E}_P(\mathbf{1}_{\{V(T)<F\}}V^2(T)) + F^2 P(V(T) \geq F)$$
$$= V(0)^2 e^{2(\mu-\frac{1}{2}\sigma^2)T} \int_{-\infty}^d \frac{1}{\sqrt{2\pi}} e^{-\frac{1}{2}x^2} e^{2\sigma\sqrt{T}x} dx + F^2(1 - N(d))$$
$$= V(0)^2 e^{2(\mu-\frac{1}{2}\sigma^2)T} e^{2\sigma^2 T} N(d - 2\sigma\sqrt{T}) + F^2(1 - N(d)).$$

This allows us to compute the standard deviation of the return on debt as

$$\sigma_D^2 = \mathbb{E}_P\left(\left(\frac{D(T) - D(0)}{D(0)}\right)^2\right) - \mu_D^2$$
$$= \frac{\mathbb{E}_P(D(T)^2)}{D(0)^2} - \frac{2\mathbb{E}_P(D(T))}{D(0)} + 1 - \mu_D^2,$$

where $\mathbb{E}_P(D(T)) = D(0)(1 + \mu_D)$ with μ_D computed as before.

Example 1.12
With the data from Example 1.9 we obtain $\sigma_E = 67.65\%$ and $\sigma_D = 1.29\%$.

Exercise 1.4 Compute σ_E and σ_D for the data in Example 1.9 but with 40% and 60% financing by equity.

Exercise 1.5 Compute the correlation coefficient ρ_{ED} between the returns on debt and equity by using the formula

$$\sigma_V^2 = w_E^2 \sigma_E^2 + w_D^2 \sigma_D^2 + 2w_E w_D \sigma_E \sigma_D \rho_{ED}$$

from portfolio theory (see [PTRM]), for various levels of debt, within the data of Example 1.9.

Example 1.13
Figure 1.2 shows the dependence of σ_D and σ_E on the financing structure of the company as represented by the equity ratio w_E. The graphs use the same

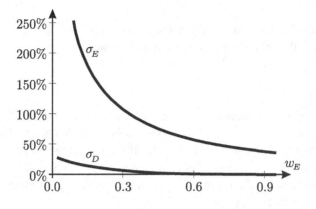

Figure 1.2 Standard deviations σ_E and σ_D as functions of equity ratio w_E.

numerical data as in Example 1.12. If $w_E = 1$, we have $\sigma_E = \sigma_V = 33.92\%$ and σ_D is undefined. If $w_E = 0$, then $\sigma_D = \sigma_V$ and σ_E makes no sense. The risk for equity is very high for equity financing below 50%, while the risk for the debt holders is also substantial in this region.

Example 1.14
High risk may be acceptable if it is accompanied by high expected return. The relationship between return and risk is captured by the market price of risk, which is a convenient way to quantify the additional return over and above the risk-free return to compensate for risk. It is given by $\frac{\mu_E - R}{\sigma_E}$ and $\frac{\mu_D - R}{\sigma_D}$ for equity and debt, where $1 + R = e^{rT}$. Maximising the market price of risk is a possible goal for the investors. In the case considered in the above examples, the market price of risk grows with the corresponding ratio:

w_E	0.1	0.2	0.3	0.4	0.5	0.6	0.7
$\frac{\mu_E - R}{\sigma_E}$	12.30%	14.31%	15.24%	15.68%	15.85%	15.89%	15.89%
w_D	0.9	0.8	0.7	0.6	0.5	0.4	0.3
$\frac{\mu_D - R}{\sigma_D}$	15.63%	13.68%	11.05%	7.94%	4.68%	1.95%	0.42%

The point where this quantity is the same for both participants may be regarded as an equilibrium of some kind. In the case in hand this takes place at $w_D = 0.8172$, a debt ratio which may be rejected by a conservative debt holder. This would correspond to a risky bond with a high return and high probability of default, a so-called junk bond.

Remark 1.15

When a company's income is taxed, interest on debt is a tax-deductible expense. This is an advantage for equity holders, called a **tax shield**, but it makes the analysis more complicated. If tax is charged at a rate R_{tax} on the difference between the company's income $V(T) - V(0)$ and the interest $F - D(0)$ whenever the resulting difference is positive, then equity holders will be left with the payoff

$$E(T) = (V(T) - R_{tax}(V(T) - V(0) - F + D(0))^+ - F)^+,$$

after the debt and tax are settled. In particular, they walk away with nothing if the company's value $V(T)$ turns out to be insufficient to pay both the debt and the tax.

Credit spread

In the literature on credit risk the financial structure of a company is often described in terms of a ratio related to F, the amount due, rather than to the initial debt value $D(0)$. This is in line with the point of view that F is the principal parameter. Compared with Definition 1.4, in the formula below $D(0)$ is replaced by the present value of F computed using the risk-free rate. The result is an abstract quantity, which nonetheless becomes a handy tool when analysing the credit spread.

Definition 1.16

The **quasi debt ratio** is defined as

$$l = \frac{Fe^{-rT}}{V(0)}.$$

Note that $w_D < l$ since $k_D > r$. The role of l can be seen if we use the Black–Scholes formula to compute the credit spread.

Proposition 1.17

The credit spread is given by

$$s = -\frac{1}{T} \ln\left(\frac{N(-d_+)}{l} + N(d_-)\right),$$

where d_+, d_- are as in (1.2). *To highlight the dependence on l we write the expressions for* d_+, d_- *as*

$$d_+ = \frac{-\ln l + \frac{1}{2}\sigma^2 T}{\sigma \sqrt{T}}, \quad d_- = \frac{-\ln l - \frac{1}{2}\sigma^2 T}{\sigma \sqrt{T}}.$$

Proof   In the proof of Proposition 1.2, formula (1.1), we saw that

$$D(0) = Fe^{-rT} - P(0),$$

where $P(0)$ is the price of a put option with strike F and expiry time T written on the company value $V(T)$. Recall the Black–Scholes formula for the put price (see [BSM])

$$P(0) = -V(0)N(-d_+) + Fe^{-rT}N(-d_-).$$

It follows that

$$s = k_D - r = \frac{1}{T}\ln\left(\frac{F}{D(0)}\right) - r$$

$$= -\frac{1}{T}\ln\left(\frac{D(0)}{Fe^{-rT}}\right) = -\frac{1}{T}\ln\left(\frac{Fe^{-rT} - P(0)}{Fe^{-rT}}\right)$$

$$= -\frac{1}{T}\ln\left(1 - \frac{-V(0)N(-d_+) + Fe^{-rT}N(-d_-)}{Fe^{-rT}}\right)$$

$$= -\frac{1}{T}\ln\left(1 + \frac{N(-d_+)}{l} - N(-d_-)\right).$$

To conclude, it is sufficient to note that $1 - N(-d_-) = N(d_-)$. $\qquad\square$

Example 1.18
Figure 1.3 shows the credit spread term structure (i.e. s as a function of debt maturity T) for $\sigma = 30\%$ and for various levels of the quasi debt ratio, from $l = 0.5$ to 1.1. We can see that the credit spread converges to zero as $T \searrow 0$ when $l < 1$ and to ∞ when $l \geq 1$. This fact can be proved.

Proposition 1.19
Write

$$s(T) = -\frac{1}{T}\ln\left(\frac{N(-d_+(T))}{l} + N(d_-(T))\right)$$

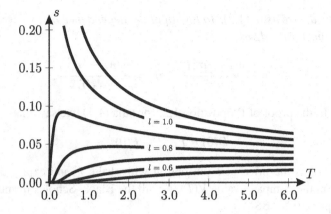

Figure 1.3 Credit spread term structure in Merton's model for various quasi debt ratios *l*.

with

$$d_+(T) = \frac{-\ln l + \frac{1}{2}\sigma^2 T}{\sigma\sqrt{T}}, \quad d_-(T) = \frac{-\ln l - \frac{1}{2}\sigma^2 T}{\sigma\sqrt{T}}.$$

Then

$$\lim_{T\searrow 0} s(T) = \begin{cases} 0 & \text{when } l < 1, \\ \infty & \text{when } l \geq 1. \end{cases}$$

Proof Let us consider the case when $l < 1$. Then

$$\lim_{T\searrow 0} d_+(T) = \lim_{T\searrow 0} \frac{-\ln l + \frac{1}{2}\sigma^2 T}{\sigma\sqrt{T}} = \infty,$$

$$\lim_{T\searrow 0} d_-(T) = \lim_{T\searrow 0} \frac{-\ln l - \frac{1}{2}\sigma^2 T}{\sigma\sqrt{T}} = \infty.$$

As a result,

$$\lim_{T\searrow 0}\left(\frac{1}{l}N(-d_+(T)) + N(d_-(T))\right) = \frac{1}{l}\lim_{x\to\infty} N(-x) + \lim_{x\to\infty} N(x) = 1$$

and

$$\lim_{T\searrow 0} \ln\left(\frac{1}{l}N(-d_+(T)) + N(d_-(T))\right) = 0.$$

It follows that we can apply l'Hôpital's rule to compute the limit

$$\lim_{T \searrow 0} s(T) = -\lim_{T \searrow 0} \frac{\frac{d}{dT} \ln\left(\frac{1}{l}N(-d_+(T)) + N(d_-(T))\right)}{\frac{d}{dT}(T)}$$

$$= -\lim_{T \searrow 0} \frac{\frac{d}{dT}\left(\frac{1}{l}N(-d_+(T)) + N(d_-(T))\right)}{\frac{1}{l}N(-d_+(T)) + N(d_-(T))}$$

$$= -\frac{1}{l}\lim_{T \searrow 0} \frac{d}{dT}\left(N(-d_+(T))\right) - \lim_{T \searrow 0} \frac{d}{dT}\left(N(d_-(T))\right)$$

$$= \frac{1}{l\sqrt{2\pi}}\lim_{T \searrow 0} e^{-\frac{1}{2}d_+(T)^2} \frac{d}{dT}(d_+(T)) - \frac{1}{\sqrt{2\pi}}\lim_{T \searrow 0} e^{-\frac{1}{2}d_-(T)^2} \frac{d}{dT}(d_-(T))$$

$$= 0$$

since the exponentials $e^{-\frac{1}{2}d_+(T)^2}$ and $e^{-\frac{1}{2}d_-(T)^2}$ converge to zero faster than the derivatives

$$\frac{d}{dT}(d_+(T)) = \frac{T\sigma^2 + 2\ln l}{4T^{\frac{3}{2}}\sigma}, \qquad \frac{d}{dT}(d_-(T)) = -\frac{T\sigma^2 - 2\ln l}{4T^{\frac{3}{2}}\sigma}$$

converge to ∞ or $-\infty$ as $T \searrow 0$.

The case when $l \geq 1$ is covered in Exercise 1.6. $\square$

Exercise 1.6   Complete the proof of Proposition 1.19 for $l \geq 1$.

Remark 1.20

The fact that in the economically realistic case when $l < 1$ the credit spread s vanishes in the short term is regarded as a major drawback of Merton's structural model. It is inconsistent with empirical studies, which show that the model underestimates such spreads.

Fixed assets

The assumption that all assets are traded is unrealistic for a typical company. In Section 1.3 we are going to relax it, while here we present a version pointing in this direction. We are going to modify the previous model, where the company's activity is based on trading stock, by including some fixed assets needed to run the company, such as office equipment, computers, and the like. These fixed assets are not actively traded, but have market value, which we denote by L and assume to be constant for simplicity. In the case of a shortage of funds they can be sold to settle the obligations. Therefore we have

$$V(t) = L + S(t),$$

where $S(t)$ is the tradeable asset (stock) price, which follows the log-normal process

$$dS(t) = rS(t)dt + \sigma S(t)dW_Q(t).$$

The company is financed by equity and debt as before,

$$V(0) = E(0) + D(0),$$

and we seek $F = D(0)e^{k_D T}$ for a given debt level $D(0)$. The equity and debt payoffs at time T are as follows:

- If $S(T) \geq F$, then debt is fully settled, $D(T) = F$, the fixed assets remain intact, and the value of equity becomes $E(T) = L + S(T) - F$.
- If $S(T) < F$ and $L + S(T) \geq F$, then $D(T) = F$. All traded assets and some of the fixed assets have to be sold. The equity value $E(T) = L + S(T) - F$ is what remains of the fixed assets.
- If $L + S(T) < F$, then $D(T) = L + S(T)$ and $E(T) = 0$. All assets are sold and bankruptcy is declared.

It means that equity is a call option on $S(T)$ with strike price $F - L$ (rather than on $V(T)$ with strike price F as before), whose payoff is

$$E(T) = \max\{S(T) - (F - L), 0\},$$

while the debt payoff is

$$D(T) = \min\{F, L + S(T)\} = F - \max\{(F - L) - S(T), 0\}.$$

Clearly,

$$E(T) + D(T) = L + S(T) = V(T).$$

The value of F can be found by matching the given initial debt level $D(0)$ to the time 0 value of the debt payoff $D(T)$, that is,

$$D(0) = e^{-rT}\mathbb{E}_Q(\min\{F, L + S(T)\}).$$

Example 1.21
With the data from Example 1.9, take $L = 5$. For debt financing at $w_D = 70\%$ we have $\mu_E = 20.89\%$, $\mu_D = 5.32\%$, and the credit spread is $s = 1.54\%$.

1.3 Non-tradeable assets

We move on to consider a company using specialised equipment or licences which cannot be sold, in addition to fixed assets (office equipment, computer hardware, etc.) which can. The investment in such a company is partially irreversible.

At time 0 the assets are purchased for

$$V(0) = L + G(0),$$

where L is the value of the fixed assets and $G(0)$ is the amount invested to acquire the non-tradeable assets. The fixed assets are liquid and can be sold for L at any time. This action may be necessary, in particular, to meet the debt obligations of the company. The value $G(0)$ is real as some licences or specialised machinery are purchased at time 0. However, it will not be possible to sell these assets, so their future value $G(t)$ becomes nominal, determined by the investors' belief that the company will generate profits. For this reason, the term **goodwill** is used when referring to $G(t)$. Goodwill is a valuable asset in its own right, and its dynamics will be modelled by a log-normal process as

$$dG(t) = \mu G(t)dt + \sigma G(t)dW_P(t),$$

where $W_P(t)$ is a Wiener process under the real-life probability P.

The profit will be described by a process $\Pi(t)$, starting from $\Pi(0) = 0$ and not necessarily positive. We make a simple assumption that increments in goodwill are proportional to the increments in profits,

$$G(t) - G(0) = q(\Pi(t) - \Pi(0)).$$

This gives

$$G(t) = q\Pi(t) + G(0).$$

In contrast to goodwill $G(t)$, the profit $\Pi(t)$ can be observed (or at least predicted in the business plan), and this relationship between $G(t)$ and $\Pi(t)$ makes it possible to calibrate the parameters μ, σ, q.

As before, the purchase of assets at time 0 is financed by equity and debt,

$$V(0) = E(0) + D(0).$$

The amount due to settle the debt at time T together with interest at a rate k_D (continuously compounded) is

$$F = e^{k_D T}D(0).$$

At time T the amount of cash available to the company to settle its debt will be $L + \Pi(T)$. The debt payoff is as follows:

- If $L + \Pi(T) \geq F$, then the debt holders will receive F, though the company may need to sell some fixed assets when the profit $\Pi(T)$ is insufficient to cover F.
- If $L + \Pi(T) < F$, then there is not enough cash to settle the debt, and bankruptcy is declared. The debt holders then receive $L + \Pi(T)$ if this amount is positive. If negative, they are not responsible for the shortfall and simply receive nothing.

Hence, the debt payoff is

$$D(T) = F\mathbf{1}_{\{L+\Pi(T)\geq F\}} + (L + \Pi(T))\,\mathbf{1}_{\{0\leq L+\Pi(T)<F\}}.$$

The equity payoff $E(T)$ is a nominal rather than actual value because it depends on the non-tradeable goodwill value $G(T)$ reflecting the company's potential to generate profits beyond time T.

Debt pricing

When the financial structure of the company is given, that is, the initial debt $D(0)$ is known, the corresponding value of F and hence the loan rate k_D can be computed so that the time 0 value of the debt payoff $D(T)$ matches the initial debt $D(0)$.

The debt payoff depends on goodwill $G(t)$ via the profit $\Pi(t)$. Because goodwill is non-tradeable, it cannot be used to replicate the payoff, which means that risk-neutral valuation is not applicable. Instead, the expected return on debt μ_D can be used to express the initial value of debt $D(0)$ in terms of the expected debt payoff $D(T)$ under the real-life probability P as

$$D(0) = \frac{\mathbb{E}_P(D(T))}{1 + \mu_D}.$$

The expectation $\mathbb{E}_P(D(T))$ is computed below, but the value of μ_D is also needed before this formula for $D(0)$ can be applied. Here we assume that the rate μ_D will be set at a suitable level above the risk-free return $R = e^{rT} - 1$ to provide compensation for risk at a given level of market price of risk

$$p = \frac{\mu_D - R}{\sigma_D},$$

where

$$\sigma_D = \frac{1}{D(0)} \sqrt{\mathbb{E}_P\left(D(T)^2\right) - \mathbb{E}_P\left(D(T)\right)^2} = \frac{1}{D(0)} \sqrt{\mathrm{Var}_P(D(T))}.$$

is the standard deviation of the return $\frac{D(T)-D(0)}{D(0)}$ on debt. This gives

$$\mu_D = p\sigma_D + R,$$

and implies that

$$D(0) = \frac{\mathbb{E}_P(D(T))}{1+\mu_D} = \frac{\mathbb{E}_P(D(T))}{1+p\sigma_D+R} = \frac{D(0)\mathbb{E}_P(D(T))}{p\sqrt{\text{Var}_P(D(T))}+D(0)(1+R)},$$

so

$$D(0) = \frac{\mathbb{E}_P(D(T)) - p\sqrt{\text{Var}_P(D(T))}}{1+R}.$$

Now we compute the expectation $\mathbb{E}_P(D(T))$ of the debt payoff under the real-life probability P. The variance $\text{Var}_P(D(T))$ is computed in Exercise 1.8. First we write

$$\mathbb{E}_P(D(T)) = \mathbb{E}_P\big(F\mathbf{1}_{\{L+\Pi(T)\geq F\}} + (L+\Pi(T))\,\mathbf{1}_{\{0\leq L+\Pi(T)<F\}}\big).$$

Here

$$\Pi(T) = \frac{1}{q}(G(T)-G(0))$$

and

$$G(T) = G(0)e^{(\mu-\frac{1}{2}\sigma^2)T+\sigma\sqrt{T}X},$$

where X is a random variable with the standard normal distribution $N(0,1)$ under the real-life probability P. When $-qL+G(0)>0$, we have

$$\mathbb{E}_P(D(T)) = FP(X\geq a_2) + \left(L-\frac{1}{q}G(0)\right)P(a_1\leq X<a_2)$$

$$+\frac{1}{q}G(0)\mathbb{E}_P\big(e^{(\mu-\frac{1}{2}\sigma^2)T+\sigma\sqrt{T}X}\mathbf{1}_{\{a_1\leq X<a_2\}}\big)$$

$$= FN(-a_2) + \left(L-\frac{1}{q}G(0)\right)(N(a_2)-N(a_1))$$

$$+\frac{1}{q}G(0)e^{\mu T}\left(N\big(a_2-\sigma\sqrt{T}\big) - N\big(a_1-\sigma\sqrt{T}\big)\right), \qquad (1.4)$$

where

$$a_1 = \frac{\ln\left(\frac{-qL+G(0)}{G(0)}\right) - \left(\mu-\frac{1}{2}\sigma^2\right)T}{\sigma\sqrt{T}}, \qquad (1.5)$$

$$a_2 = \frac{\ln\left(\frac{qF-qL+G(0)}{G(0)}\right) - \left(\mu-\frac{1}{2}\sigma^2\right)T}{\sigma\sqrt{T}}. \qquad (1.6)$$

When $-qL + G(0) \leq 0$, the inequality $0 \leq L + \Pi(T)$ is always satisfied, and we get

$$\mathbb{E}_P(D(T)) = FP(X \geq a_2) + \left(L - \frac{1}{q}G(0)\right)P(X < a_2)$$

$$+ \frac{1}{q}G(0)\mathbb{E}_P\left(e^{(\mu - \frac{1}{2}\sigma^2)T + \sigma\sqrt{T}X}\mathbf{1}_{\{X < a_2\}}\right)$$

$$= FN(-a_2) + \left(L - \frac{1}{q}G(0)\right)N(a_2)$$

$$+ \frac{1}{q}G(0)e^{\mu T}N\left(a_2 - \sigma\sqrt{T}\right). \tag{1.7}$$

Exercise 1.7 Show that the probability of default is $N(a_2)$.

Exercise 1.8 Derive a formula for $\mathbb{E}_P(D(T)^2)$ and hence for the variance $\mathrm{Var}_P(D(T))$ in terms of the standard normal distribution function $N(x)$.

Example 1.22
Consider a company with $L = 50$ invested into fixed assets and $G(0) = 50$ into non-tradeable assets, so $V(0) = L + G(0) = 100$. Suppose that $\mu = 10\%$, $\sigma = 30\%$, $q = 1$, and the market price of risk is $p = 0.82$. Assuming debt financing at $w_D = 25\%$, the amount borrowed is $D(0) = 25$. For debt maturity $T = 1$ we compute the value $F = 26.5908$ that gives this debt level. The corresponding loan rate is $k_D = 6.17\%$.

1.4 Barrier model

If the debt is in the form of a bond and the company's performance is poor prior to the debt maturity time T, this will be reflected in the bond prices. The bond holders would then have a chance to sell the bond early. For a bank loan the situation is different. Early termination of the debt will be possible only if a suitable clause is included in the debt specifications.

The agreement between the bank and the company has to include two items:

- the criterion for an early default,
- the payoff to the bank.

In order to formulate a default criterion, in this section we again assume that the company's assets are tradeable and denote their value by $V(t)$. Bankruptcy is declared as soon as $V(t)$ reaches or drops below a given deterministic function $B(t)$, called a **barrier**. (In practice, $V(t)$ will be replaced by a proxy based on some financial ratios describing the condition of the company.) The time of default becomes a random variable τ defined as

$$\tau = \inf\{t \in [0, T] : V(t) \le B(t)\},$$

with $\inf \varnothing = \infty$. To exclude the case $\tau = 0$, we assume that $B(0) < V(0)$. Assume for simplicity that the paths of $V(t)$ and the function $B(t)$ are continuous. Then $\tau = \inf\{t \in [0, T] : V(t) = B(t)\}$ and $V(\tau) = B(\tau)$.

The bank will receive an agreed sum F if the company stays solvent in the whole period $[0, T]$. We assume that in the case of bankruptcy the bank payment never exceeds F, of course, and it is strictly smaller than F with positive probability (to avoid the trivial case of riskless debt). When $T < \tau$, the payoff to the bank is F if $V(T) \ge F$ and $V(T)$ otherwise, so it is given by $\min\{F, V(T)\}$, like in the Merton model. If the barrier is reached prior to or at maturity, that is, if $\tau \le T$, then the bank will receive the amount $V(\tau)$ at time τ, or equivalently $V(\tau)e^{r(T-\tau)}$ at time T.

In order to be able to derive an explicit formula for the debt value, we assume a simple form of the barrier:

$$B(t) = Fe^{-\gamma(T-t)},$$

the value of γ to be specified in the debt agreement. Note that, since $B(T) = F$, the bank receives F if the company survives, that is, $\tau > T$. In the case of bankruptcy, $\tau \le T$, the assumption $V(\tau)e^{r(T-\tau)} = B(\tau)e^{r(T-\tau)} < F$ reads $e^{(r-\gamma)(T-\tau)} < 1$, implying the condition $\gamma > r$, which we now impose.

Finally, the assumption $B(0) < V(0)$ implies that

$$L = \frac{Fe^{-\gamma T}}{V(0)} < 1.$$

The number L will play an important role in what follows. It looks similar to the quasi debt ratio discussed in Section 1.2, but with r replaced by γ. The relationship between the value and the barrier can be analysed by considering their ratio $\frac{Fe^{-\gamma(T-t)}}{V(t)}$, which starts from L, and default happens when it hits 1.

Remark 1.23
Since the paths of $V(t)$ are continuous, τ is a so-called **predictable stopping time**. By definition, it means that there exists a sequence of stopping times τ_n for $n = 1, 2, \ldots$ such that, almost surely, $\tau_n \leq \tau_{n+1}$ for all n, $\lim_{n\to\infty} \tau_n = \tau$, and if $\tau < \infty$ then $\tau_n < \tau$ for all n. Note that all stopping times are considered with respect to the filtration generated by $V(t), t \geq 0$; see [SCF]. Such a sequence of stopping times can be constructed as

$$\tau_n = \inf\{t \in [0, T] : V(t) \leq Fe^{-\gamma(T-t)} + \tfrac{1}{n}\}. \tag{1.8}$$

Predictability is regarded as a weakness of the barrier model because in real life a default can sometimes happen without any prior warning.

Exercise 1.9   Prove that the sequence τ_n defined in (1.8) satisfies the conditions for τ to be a predictable stopping time.

We work with the assumptions of the Black–Scholes model, hence

$$dV(t) = \mu V(t)dt + \sigma V(t)dW_P(t),$$

where W_P is a Wiener process under the real-life probability P. By applying Girsanov's theorem, we obtain

$$dV(t) = rV(t)dt + \sigma V(t)dW_Q(t),$$

where W_Q is a Wiener process under the risk-neutral probability Q.
  The equity and debt payoffs at time T are

$$E(T) = (V(T) - F)\mathbf{1}_{\{\tau>T\}},$$
$$D(T) = F\mathbf{1}_{\{\tau>T\}} + Fe^{(r-\gamma)(T-\tau)}\mathbf{1}_{\{\tau\leq T\}}.$$

In the case of an early default at time $\tau \leq T$ the debt payoff $Fe^{-\gamma(T-\tau)}$ is invested risk free, so it becomes $Fe^{-\gamma(T-\tau)}e^{r(T-t)} = Fe^{(r-\gamma)(T-\tau)}$ at time T. Note that $E(T) + D(T)$ is not equal to $V(T)$ (as was the case in the Merton model).

Remark 1.24
The case with a barrier of the form $Ke^{-\gamma(T-t)}$, where $K \leq F$, is known as the **Black–Cox model**. It is slightly more general than the case with $K = F$ studied here, and can be analysed in a similar manner. The main difference

is that default may take place at time T even if $\tau > T$. The payoffs for equity and debt become

$$E(T) = (V(T) - F)\mathbf{1}_{\{\tau \geq T, V(T) \geq F\}},$$
$$D(T) = F\mathbf{1}_{\{\tau \geq T, V(T) \geq F\}} + V(T)\mathbf{1}_{\{\tau \geq T, V(T) < F\}} + Ke^{(r-\gamma)(T-\tau)}\mathbf{1}_{\{\tau < T\}}.$$

Proposition 1.25
Let F_M and, respectively, F_B denote the debt repayment amount in the Merton and barrier models, both corresponding to the same initial equity value $E(0)$. Then $F_B \leq F_M$.

Proof The initial equity value is equal to the discounted expected payoff under the risk-neutral probability, so

$$E(0) = e^{-rT}\mathbb{E}_Q((V(T) - F_M)\mathbf{1}_{\{V(T) > F_M\}})$$

in the Merton model, and

$$E(0) = e^{-rT}\mathbb{E}_Q((V(T) - F_B)\mathbf{1}_{\{\tau > T\}})$$

in the barrier model. It follows that

$$\mathbb{E}_Q((V(T) - F_M)\mathbf{1}_{\{V(T) > F_M\}}) = \mathbb{E}_Q((V(T) - F_B)\mathbf{1}_{\{\tau > T\}})$$
$$\leq \mathbb{E}_Q((V(T) - F_B)\mathbf{1}_{\{V(T) > F_B\}}),$$

where the inequality holds because $\{\tau > T\} \subset \{V(T) > F_B\}$.

Now suppose that $F_B > F_M$. Then we would have

$$V(T) - F_M > V(T) - F_B$$

and

$$\mathbf{1}_{\{V(T) > F_M\}} \geq \mathbf{1}_{\{V(T) > F_B\}},$$

hence

$$\mathbb{E}_Q((V(T) - F_M)\mathbf{1}_{\{V(T) > F_M\}}) > \mathbb{E}_Q((V(T) - F_B)\mathbf{1}_{\{V(T) > F_B\}}),$$

contradicting the above inequality. This proves that $F_B \leq F_M$. $\qquad\square$

Corollary 1.26
The credit spread s_B in the barrier model cannot exceed the credit spread s_M in the Merton model, that is, $s_B \leq s_M$.

Proof The inequality $F_B \leq F_M$ implies the corresponding inequality for $s_B = \frac{1}{T}\ln\frac{F_B}{D(0)} - r$ and $s_M = \frac{1}{T}\ln\frac{F_M}{D(0)} - r$. $\qquad\square$

Exercise 1.10   Verify the inequality $F_B \leq F_M$ by using the debt payoff instead of the equity payoff.

For the initial equity and debt values we obtain the equations

$$E(0) = e^{-rT}\mathbb{E}_Q((V(T) - F)\mathbf{1}_{\{\tau > T\}}),$$
$$D(0) = e^{-rT}\mathbb{E}_Q(F\mathbf{1}_{\{\tau > T\}} + Fe^{(r-\gamma)(T-\tau)}\mathbf{1}_{\{\tau \leq T\}}),$$

with two variables F and γ (both also implicitly contained in τ). As we
shall see in Proposition 1.31, the balance sheet equation $E(0)+D(0) = V(0)$
holds for every F and γ, so we have in fact just a single equation with two
variables and possibly infinitely many solutions.

Auxiliary facts

It will be convenient to describe τ in terms of the process

$$Y(t) = -\ln \frac{Fe^{-\gamma(T-t)}}{V(t)}$$
$$= -\ln \frac{Fe^{-\gamma T}}{V(0)} + \left(r - \gamma - \frac{1}{2}\sigma^2\right)t + \sigma W_Q(t)$$
$$= -\ln L + \left(r - \gamma - \frac{1}{2}\sigma^2\right)t + \sigma W_Q(t).$$

We can express τ as

$$\tau = \inf\{t \in [0,T] : V(t) \leq Fe^{-\gamma(T-t)}\}$$
$$= \inf\{t \in [0,T] : Y(t) \leq 0\}.$$

Next, we find the cumulative distribution function of τ under the risk-
neutral probability Q, which will be needed to compute certain expectations involving τ.

Lemma 1.27
For each $t > 0$,

$$Q(\tau \leq t) = N(d_1(t)) + L^{2\alpha}N(d_2(t)),$$

where

$$\alpha = \frac{r - \gamma - \frac{1}{2}\sigma^2}{\sigma^2} < 0,$$

$$d_1(t) = \frac{\ln L - \left(r - \gamma - \frac{1}{2}\sigma^2\right)t}{\sigma\sqrt{t}},$$

$$d_2(t) = \frac{\ln L + \left(r - \gamma - \frac{1}{2}\sigma^2\right)t}{\sigma\sqrt{t}}.$$

Proof This follows directly from Proposition A.17, which for $Y(t) = y_0 + vt + \sigma W_Q(t)$ with $\tau = \inf\{t \geq 0 : Y(t) \leq 0\}$ gives

$$Q(\tau \leq t) = N\left(\frac{-y_0 - vt}{\sigma\sqrt{t}}\right) + e^{-2v\sigma^{-2}y_0}N\left(\frac{-y_0 + vt}{\sigma\sqrt{t}}\right).$$

All that is needed is to insert $y_0 = -\ln L$ and $v = r - \gamma - \frac{1}{2}\sigma^2$ and to observe that

$$e^{-2v\sigma^{-2}y_0} = e^{2\sigma^{-2}\left(r - \gamma - \frac{1}{2}\sigma^2\right)\ln L} = L^{2\alpha}.$$

$\square$

We also need the following result.

Lemma 1.28
For any $x > F$,

$$Q(V(T) \geq x, \tau \geq T)$$
$$= N\left(\frac{\ln\frac{V(0)}{x} + \left(r - \frac{1}{2}\sigma^2\right)T}{\sigma\sqrt{T}}\right) - L^{2\alpha}N\left(\frac{\ln\frac{F^2 e^{-2\gamma T}}{xV(0)} + \left(r - \frac{1}{2}\sigma^2\right)T}{\sigma\sqrt{T}}\right),$$

with α given in Lemma 1.27.

Proof From Proposition A.18 applied to $Y(t) = y_0 + vt + \sigma W_Q(t)$ we have

$$Q(Y(T) \geq y, \tau \geq T) = N\left(\frac{-y + y_0 + vT}{\sigma\sqrt{T}}\right) - e^{-2v\sigma^{-2}y_0}N\left(\frac{-y - y_0 + vT}{\sigma\sqrt{T}}\right).$$

We use it with

$$y = \ln\frac{x}{F},$$

$$y_0 = -\ln L = -\ln\frac{Fe^{-\gamma T}}{V(0)},$$

$$v = r - \gamma - \frac{1}{2}\sigma^2,$$

and observe that $V(T) \geq x$ if and only if $Y(T) \geq y$. This gives

$$Q(V(T) \geq x, \tau \geq T)$$
$$= Q(Y(T) \geq y, \tau \geq T)$$
$$= N\left(\frac{-y + y_0 + vT}{\sigma\sqrt{T}}\right) - e^{-2v\sigma^{-2}y_0} N\left(\frac{-y - y_0 + vT}{\sigma\sqrt{T}}\right)$$
$$= N\left(\frac{-\ln\frac{x}{F} - \ln\frac{Fe^{-\gamma T}}{V(0)} + \left(r - \gamma - \frac{1}{2}\sigma^2\right)T}{\sigma\sqrt{T}}\right)$$
$$\quad - e^{-2v\sigma^{-2}y_0} N\left(\frac{-\ln\frac{x}{F} + \ln\frac{Fe^{-\gamma T}}{V(0)} + \left(r - \gamma - \frac{1}{2}\sigma^2\right)T}{\sigma\sqrt{T}}\right)$$
$$= N\left(\frac{\ln\frac{V(0)}{x} + \left(r - \frac{1}{2}\sigma^2\right)T}{\sigma\sqrt{T}}\right) - L^{2\alpha}N\left(\frac{\ln\frac{F^2 e^{-2\gamma T}}{xV(0)} + \left(r - \frac{1}{2}\sigma^2\right)T}{\sigma\sqrt{T}}\right).$$

$\square$

Debt

Theorem 1.29
The initial value of debt has the form

$$D(0) = Fe^{-rT}\left(N(-d_1) - L^{2\alpha}N(d_2)\right) + V(0)\left(N(d_3) + L^{2\alpha+2}N(d_4)\right),$$

with α, $d_1 = d_1(T)$ and $d_2 = d_2(T)$ given in Lemma 1.27, and with

$$d_3 = \frac{\ln L + \sigma\beta T}{\sigma\sqrt{T}},$$
$$d_4 = \frac{\ln L - \sigma\beta T}{\sigma\sqrt{T}},$$

where

$$\beta = -\sigma(\alpha + 1).$$

Proof We need to compute the two terms on the right-hand side of the pricing formula

$$D(0) = e^{-rT}\mathbb{E}_Q(F1_{\{\tau > T\}}) + e^{-rT}\mathbb{E}_Q(Fe^{(r-\gamma)(T-\tau)}1_{\{\tau \leq T\}}).$$

The first term is concerned with the terminal payoff, which happens when the barrier is not hit. Lemma 1.27 gives a formula for $Q(\tau \leq T)$. Since

$$\mathbb{E}_Q(1_{\{\tau > T\}}) = 1 - Q(\tau \leq T),$$

all we need is to notice that $1 - N(a) = N(-a)$ to get

$$e^{-rT}\mathbb{E}_Q(F\mathbf{1}_{\{\tau>T\}}) = Fe^{-rT}\left(N(-d_1) - L^{2\alpha}N(d_2)\right).$$

The other term, related to bankruptcy caused by hitting the barrier, is equal to

$$e^{-rT}\mathbb{E}_Q(Fe^{(r-\gamma)(T-\tau)}\mathbf{1}_{\{\tau\leq T\}}) = Fe^{-\gamma T}\mathbb{E}_Q(e^{(\gamma-r)\tau}\mathbf{1}_{\{\tau\leq T\}})$$
$$= V(0)L\mathbb{E}_Q(e^{(\gamma-r)\tau}\mathbf{1}_{\{\tau\leq T\}}),$$

so we need to prove that

$$\mathbb{E}_Q(e^{(\gamma-r)\tau}\mathbf{1}_{\{\tau\leq T\}}) = L^{-1}N(d_3) + L^{2\alpha+1}N(d_4). \tag{1.9}$$

We are going to verify this equality in the case when $\gamma \neq r + \frac{1}{2}\sigma^2$. The case of $\gamma = r + \frac{1}{2}\sigma^2$ is covered in Exercise 1.11. To compute the expectation we use the distribution function of τ in Lemma 1.27, which we write as

$$Q(\tau \leq t) = N\left(\frac{-a - b_1 t}{\sqrt{t}}\right) + L^{2\alpha}N\left(\frac{-a - b_2 t}{\sqrt{t}}\right),$$

with constants

$$a = -\frac{1}{\sigma}\ln L,$$

$$b_1 = -b_2 = \frac{1}{\sigma}\left(r - \gamma - \frac{1}{2}\sigma^2\right),$$

where $a > 0$ since $L < 1$. Hence

$$\mathbb{E}_Q(e^{(\gamma-r)\tau}\mathbf{1}_{\{\tau\leq T\}}) = \int_0^T e^{(\gamma-r)t}dN\left(\frac{-a - b_1 t}{\sqrt{t}}\right) + L^{2\alpha}\int_0^T e^{(\gamma-r)t}dN\left(\frac{-a - b_2 t}{\sqrt{t}}\right).$$

Integrals of this type are computed in Example A.12:

$$\int_0^y e^{cx}dN\left(\frac{-a - bx}{\sqrt{x}}\right)$$
$$= \frac{d + b}{2d}e^{-a(b-d)}N\left(\frac{-a - dy}{\sqrt{y}}\right) + \frac{d - b}{2d}e^{-a(b+d)}N\left(\frac{-a + dy}{\sqrt{y}}\right),$$

where $d = \sqrt{b^2 - 2c}$, provided that $b^2 > 2c$ and $a > 0$. Observe that the right-hand side of this formula has the same value for $d = -\sqrt{b^2 - 2c}$.

We take $y = T$, $b = b_i$, and $c = \gamma - r$. Then

$$b^2 - 2c = \frac{1}{\sigma^2}\left(r - \gamma - \frac{1}{2}\sigma^2\right)^2 + 2(r - \gamma) = \frac{1}{\sigma^2}\left(r - \gamma + \frac{1}{2}\sigma^2\right)^2 > 0$$

and

$$d = \sqrt{b^2 - 2c} = \sqrt{\frac{1}{\sigma^2}\left(r - \gamma - \frac{1}{2}\sigma^2\right)^2 + 2(r - \gamma)}.$$

Moreover,

$$\beta^2 = (-\sigma(\alpha + 1))^2$$
$$= \left(\frac{1}{\sigma}\left(r - \gamma - \frac{1}{2}\sigma^2\right) + \sigma\right)^2$$
$$= \frac{1}{\sigma^2}\left(r - \gamma - \frac{1}{2}\sigma^2\right)^2 + 2(r - \gamma)$$
$$= d^2;$$

hence

$$\beta = \begin{cases} d & \text{if } \alpha + 1 < 0, \\ -d & \text{if } \alpha + 1 > 0. \end{cases}$$

It follows that

$$\mathbb{E}_Q(e^{(\gamma - r)\tau}\mathbf{1}_{\{\tau \le T\}}) = \frac{\beta + b_1}{2\beta}e^{-a(b_1 - \beta)}N(d_4) + \frac{\beta - b_1}{2\beta}e^{-a(b_1 + \beta)}N(d_3)$$
$$+ L^{2\alpha}\frac{\beta + b_2}{2\beta}e^{-a(b_2 - \beta)}N(d_4) + L^{2\alpha}\frac{\beta - b_2}{2\beta}e^{-a(b_2 + \beta)}N(d_3),$$

in both cases when $\alpha + 1 < 0$ and $\alpha + 1 > 0$. Inserting $b_2 = -b_1$ and

$$e^{-ab_1} = e^{ab_2} = e^{\frac{1}{\sigma}(\ln L)\frac{1}{\sigma}(r - \gamma - \frac{1}{2}\sigma^2)} = L^\alpha,$$
$$e^{a\beta} = e^{-\frac{1}{\sigma}\ln L(-\sigma(\alpha+1))} = L^{\alpha+1},$$

we get

$$\mathbb{E}(e^{(\gamma - r)\tau}\mathbf{1}_{\{\tau < T\}}) = \frac{\beta + b_1}{2\beta}L^\alpha L^{\alpha+1}N(d_4) + \frac{\beta - b_1}{2\beta}L^\alpha L^{-\alpha-1}N(d_3)$$
$$+ L^{2\alpha}\frac{\beta - b_1}{2\beta}L^{-\alpha}L^{\alpha+1}N(d_4) + L^{2\alpha}\frac{\beta + b_1}{2\beta}L^{-\alpha}L^{-\alpha-1}N(d_3)$$
$$= L^{2\alpha+1}N(d_4) + L^{-1}N(d_3),$$

which completes the proof. $\square$

Exercise 1.11   Show that (1.9) holds in the case when $\gamma = r + \frac{1}{2}\sigma^2$.

Exercise 1.12   Show that $D(0)$ is decreasing as a function of γ when the parameters $F, V(0), r, \sigma, T$ are fixed.

Exercise 1.13   Show that $D(0)$ is increasing as a function of F when the parameters $\gamma, V(0), r, \sigma, T$ are fixed.

Equity

Theorem 1.30

The initial value of equity is given by

$$E(0) = V(0)\left(N(-d_3) - L^{2\alpha+2}N(d_4)\right) - e^{-rT}F\left(N(-d_1) - L^{2\alpha}N(d_2)\right),$$

where $\alpha, d_1, d_2, d_3, d_4$ are as in Lemma 1.27 and Theorem 1.29.

Proof   We need to compute the expectation

$$\begin{aligned}
E(0) &= e^{-rT}\mathbb{E}_Q((V(T) - F)\mathbf{1}_{\{\tau > T\}}) \\
&= e^{-rT}\mathbb{E}_Q(V(T)\mathbf{1}_{\{\tau > T\}}) - e^{-rT}FQ(\tau > T).
\end{aligned}$$

The second term is equal to

$$e^{-rT}FQ(\tau > T) = e^{-rT}F\left(N(-d_1) - L^{2\alpha}N(d_2)\right)$$

by Lemma 1.27 (the same argument as in the proof of Theorem 1.29). It remains to show that

$$\mathbb{E}_Q(V(T)\mathbf{1}_{\{\tau > T\}}) = e^{rT}V(0)N(-d_3) - e^{rT}V(0)L^{2\alpha+2}N(d_4).$$

By Lemma 1.28, for any $x \geq F$,

$$\begin{aligned}
&Q(V(T) > x, \tau > T) \\
&= N\left(\frac{\ln\frac{V(0)}{x} + (r - \frac{1}{2}\sigma^2)T}{\sigma\sqrt{T}}\right) - L^{2\alpha}N\left(\frac{\ln\frac{F^2 e^{-2\gamma T}}{xV(0)} + (r - \frac{1}{2}\sigma^2)T}{\sigma\sqrt{T}}\right) \\
&= 1 - F_1(x) - L^{2\alpha}F_2(x),
\end{aligned}$$

where

$$F_1(x) = N\left(\frac{\ln x - \ln V(0) - (r - \frac{1}{2}\sigma^2)T}{\sigma \sqrt{T}}\right),$$

$$F_2(x) = N\left(\frac{-\ln x - \ln V(0) + 2\ln F - 2\gamma T + (r - \frac{1}{2}\sigma^2)T}{\sigma \sqrt{T}}\right).$$

Hence

$$\mathbb{E}_Q(V(T)\mathbf{1}_{\{\tau \geq T\}}) = \int_F^\infty x dF_1(x) + L^{2\alpha} \int_F^\infty x dF_2(x).$$

We employ the formulae in Exercise A.3. First we use

$$\int_0^y x dN\left(\frac{\ln x + a}{b}\right) = e^{\frac{1}{2}b^2 - a} N\left(\frac{\ln y + a - b^2}{b}\right)$$

with $a = -\ln V(0) - (r - \frac{1}{2}\sigma^2)T$ and $b = \sigma \sqrt{T}$ to find

$$\int_F^\infty x dF_1(x) = \int_0^\infty x dF_1(x) - \int_0^F x dF_1(x)$$

$$= e^{\frac{1}{2}b^2 - a} - e^{\frac{1}{2}b^2 - a} N\left(\frac{\ln F + a - b^2}{b}\right)$$

$$= V(0)e^{rT} N\left(\frac{-\ln F + \ln V(0) + (r + \frac{1}{2}\sigma^2)T}{\sigma \sqrt{T}}\right)$$

$$= V(0)e^{rT} N(-d_3)$$

since

$$d_3 = \frac{\ln L + \sigma \beta T}{\sigma \sqrt{T}} = \frac{\ln F - \ln V(0) - (r + \frac{1}{2}\sigma^2)T}{\sigma \sqrt{T}}.$$

Next, the formula

$$\int_0^y x dN\left(\frac{-\ln x + a}{b}\right) = -e^{\frac{1}{2}b^2 + a} N\left(\frac{\ln y - a - b^2}{b}\right)$$

with $a = -\ln V(0) + 2\ln F - 2\gamma T + (r - \frac{1}{2}\sigma^2)T$ and $b = \sigma\sqrt{T}$ gives

$$\int_F^\infty x dF_2(x) = \int_0^\infty x dF_2(x) - \int_0^F x dF_2(x)$$

$$= -e^{\frac{1}{2}b^2 + a} + e^{\frac{1}{2}b^2 + a}N\left(\frac{\ln F - a - b^2}{b}\right)$$

$$= -e^{\frac{1}{2}b^2 + a}\left(1 - N\left(\frac{\ln F - a - b^2}{b}\right)\right)$$

$$= -e^{\frac{1}{2}b^2 + a}N\left(\frac{-\ln F + a + b^2}{b}\right)$$

$$= -V(0)L^2 e^{rT}N\left(\frac{\ln F - \ln V(0) - 2\gamma T + (r + \frac{1}{2}\sigma^2)T}{\sigma\sqrt{T}}\right)$$

$$= -V(0)L^2 e^{rT}N(d_4)$$

since

$$d_4 = \frac{\ln L - \sigma\beta T}{\sigma\sqrt{T}} = \frac{\ln F - \ln V(0) - 2\gamma T + (r + \frac{1}{2}\sigma^2)T}{\sigma\sqrt{T}}.$$

This completes the proof. □

The next exercise shows that the barrier becomes irrelevant as γ grows to ∞.

> **Exercise 1.14** Show that in the limit as $\gamma \to \infty$ we obtain the pricing formulae for $D(0)$ and $E(0)$ in the Merton model stated in Section 1.2.

Balance sheet equation

In the classical Merton model the time T payoffs of debt and equity add up to $V(T)$, and this implies the same relation at time 0 (by using put-call parity, for instance). In the barrier model this is more complicated, and put-call parity does not apply directly. Still, the balance sheet equation $V(0) = D(0) + E(0)$ holds at time 0.

Proposition 1.31
The balance sheet equation $V(0) = D(0) + E(0)$ holds for all γ and F.

Proof Consider the process

$$X(t) = V(t)\mathbf{1}_{\{t < \tau\}} + e^{r(t-\tau)}V(\tau)\mathbf{1}_{\{\tau \le t\}}$$

for all $t \geq 0$, and observe that

$$e^{-rt}X(t) = e^{-r(t\wedge\tau)}V(t \wedge \tau),$$

that is, the discounted process $e^{-rt}X(t)$ is equal to the discounted value process $e^{-rt}V(t)$ stopped at time τ. Because the discounted value process $e^{-rt}V(t)$ is a martingale under the risk-neutral probability Q, so is $e^{-rt}X(t)$. Moreover, since $V(\tau) = Fe^{-\gamma(T-\tau)}$,

$$
\begin{aligned}
X(T) &= V(T)\mathbf{1}_{\{T<\tau\}} + e^{r(T-\tau)}V(\tau)\mathbf{1}_{\{\tau\leq T\}} \\
&= (V(T) - F)\,\mathbf{1}_{\{T<\tau\}} + F\mathbf{1}_{\{T<\tau\}} + Fe^{(r-\gamma)(T-\tau)}\mathbf{1}_{\{\tau\leq T\}} \\
&= E(T) + D(T).
\end{aligned}
$$

It follows that

$$V(0) = X(0) = e^{-rT}\mathbb{E}_Q(X(T)) = e^{-rT}\mathbb{E}_Q(E(T) + D(T)) = E(0) + D(0).$$

$\square$

Remark 1.32
The equality $V(0) = D(0) + E(0)$ can also be verified by substituting the expressions for $D(0)$ and $E(0)$ from Theorems 1.29 and 1.30. Indeed,

$$
\begin{aligned}
D(0) &+ E(0) \\
&= Fe^{-rT}\left(N(-d_1) - L^{2\alpha}N(d_2)\right) + V(0)\left(N(d_3) + L^{2\alpha+2}N(d_4)\right) \\
&\quad + V(0)\left(N(-d_3) - L^{2\alpha+2}N(d_4)\right) - e^{-rT}F\left(N(-d_1) - L^{2\alpha}N(d_2)\right) \\
&= V(0)\,(N(d_3) + N(-d_3)) \\
&= V(0).
\end{aligned}
$$

Credit spread

Credit spread is a simple and convenient variable, which can be used to quantify the risk of default. It is defined in the same way as in the Merton model,

$$s = \frac{1}{T} \ln \frac{F}{D(0)} - r.$$

The short-term spread is zero since the spread in the barrier model cannot exceed that in the Merton model according to Corollary 1.26.

The role of the quasi debt ratio is played here by $L = \frac{Fe^{-\gamma T}}{V(0)}$, and we have

$$\frac{D(0)}{F} = e^{-rT}\left(N(-d_1) - L^{2\alpha}N(d_2)\right) + L^{-1}e^{-\gamma T}\left(N(d_3) + L^{2\alpha+2}N(d_4)\right)$$

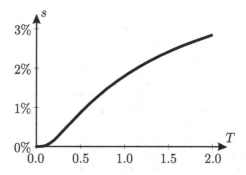

Figure 1.4 Credit spread s as a function of T in the barrier model, Example 1.33.

by Theorem 1.29. Hence we have the credit spread as a function of L only (not involving $D(0)$ or F directly):

$$s = -\frac{1}{T} \ln\left(e^{-rT}\left(N(-d_1) - L^{2\alpha}N(d_2)\right)\right.$$
$$\left. + L^{-1}e^{-\gamma T}\left(N(d_3) + L^{2\alpha+2}N(d_4)\right)\right) - r,$$

where

$$\alpha = \frac{r - \gamma - \frac{1}{2}\sigma^2}{\sigma^2} < 0$$

and

$$d_1 = \frac{\ln L - (r - \gamma - \frac{1}{2}\sigma^2)T}{\sigma\sqrt{T}},$$

$$d_2 = \frac{\ln L + (r - \gamma - \frac{1}{2}\sigma^2)T}{\sigma\sqrt{T}},$$

$$d_3 = \frac{\ln L - (r - \gamma + \frac{1}{2}\sigma^2)T}{\sigma\sqrt{T}},$$

$$d_4 = \frac{\ln L + (r - \gamma + \frac{1}{2}\sigma^2)T}{\sigma\sqrt{T}}.$$

Example 1.33

The dependence of the credit spread s on T is shown in Figure 1.4 for $r = 3\%$, $\gamma = 9\%$, $\sigma = 20\%$, and $L = 0.85$.

Table 1.1 *Credit spreads in Example 1.34.*

γ	$w_D = $ 40%	$w_D = $ 50%	$w_D = $ 60%	$w_D = $ 70%	$w_D = $ 80%	$w_D = $ 90%
10%	0.0023%	0.0273%	0.1481%	0.5051%	1.2850%	2.7014%
15%	0.0041%	0.0481%	0.2588%	0.8833%	2.2848%	4.9981%
20%	0.0056%	0.0643%	0.3430%	1.1690%	3.0564%	6.9010%
25%	0.0068%	0.0771%	0.4081%	1.3874%	3.6513%	8.4445%
30%	0.0078%	0.0874%	0.4591%	1.5564%	4.1115%	9.6775%
35%	0.0086%	0.0958%	0.4998%	1.6889%	4.4701%	10.6540%
40%	0.0093%	0.1027%	0.5325%	1.7941%	4.7520%	11.4251%
∞	0.0148%	0.1515%	0.7408%	2.3995%	6.2305%	15.1173%

Example 1.34

In Table 1.1 we list the credit spreads for various values of the debt ratio w_D and barrier parameter γ for $r = 5\%$, $\sigma = 30\%$, and $T = 1$. When w_D is under 30%, the credit spread becomes negligible, lower than 0.0004%. The last row in the table shows the credit spreads in the Merton model ($\gamma = \infty$). We can see that, by introducing a barrier with γ around 15%, debt financing can be increased by some 10% without substantially affecting the credit spread as compared to the Merton model.

Exercise 1.15 Show that the credit spread s is increasing as a function of γ when $D(0), V(0), r, \sigma, T$ are fixed.

2

Hazard function model and no arbitrage

We are going to study the so-called **reduced-form** approach as an alternative to structural models of credit risk. The default of a company will be an exogenous event, happening unexpectedly and without a direct link to the state of the company. This is a documented possibility; companies can indeed go bankrupt without warning.

A common feature in such default events is a new piece of information that is revealed, affecting the company's existence. For instance, a new scientific discovery could make the production technology obsolete or it might reduce the demand for the company's product. A court decision could damage the financial position of the company. A political event might wipe out the demand for the company's product or services in a critical region. In such cases default can happen even if the company has been in a good shape prior to the triggering event. An additional possibility is when the financial statements are doctored to cover poor performance, and the true financial state of the company suddenly becomes public knowledge.

Default will be assumed to occur at a random time τ, not directly linked to the functioning of the company, though in general not independent of it.

2.1 Market model

The model involves two types of securities, default-free and defaultable.

We begin with the simple case when there is just one default-free asset, a risk-free zero-coupon bond with maturity date $T > 0$ and unit face value, whose price at any time $t \in [0, T]$ is determined by a constant risk-free rate r as

$$B(t, T) = e^{-r(T-t)}.$$

In Chapter 5 we are going to add another default-free asset, a risky one such as some stock or an index.

In addition to the default-free bond, suppose there is a defaultable zero-coupon bond with maturity T and unit face value, issued by a company which may default in the future. Let a random variable $\tau > 0$ be the time of default. The value of the defaultable bond at time $t \in [0, T]$ will be denoted by $D(t, T)$. The bond holder will receive the face value at time T if the company does not default up to and including time T, that is, if $T < \tau$. However, if a default occurs before or at time T, that is, if $\tau \le T$, then the bond holder will receive nothing (a zero-recovery defaultable bond). Hence, at maturity T the value of the defaultable bond will be

$$D(T, T) = \begin{cases} 1 & \text{if } T < \tau, \\ 0 & \text{if } \tau \le T. \end{cases}$$

Assumption 2.1
The traded assets in the model are the risk-free zero-coupon bond $B(t, T)$ and the zero-coupon zero-recovery defaultable bond $D(t, T)$.

The next step will be to consider trading strategies constructed from $B(t, T)$ and $D(t, T)$ and see what securities can be replicated. This will allow the pricing of more complex securities, first by using the replicating strategies and ultimately by deriving formulae based on expected payoffs with respect to a certain probability measure, in line with the methods developed in the classical approaches; see [DMFM] and [BSM]. In particular, bonds with maturities shorter than T are not included among the underlying securities. They will be priced and replicated using the two underlying bonds; see Example 4.13 and Exercise 4.13.

Remark 2.2
In practice, zero-coupon zero-recovery defaultable bonds are not traded. Nonetheless, since we will be able to use them to price certain other liquid securities, the above assumption can be justified by the fact that the calibration of such bonds will follow from the market data for traded securities.

The key point here is that default takes place at a random time τ, hence we need to discuss such random times first.

2.2 Default time and hazard function

Consider a probability space $(\Omega, \mathcal{F}, P)$, which we do not specify here but assume to be sufficiently rich to accommodate all random variables considered. The probability measure P plays the role of real-life probability.

The **default time** τ is a random variable on Ω with values in $(0, \infty]$. The fact that τ is taken to be strictly positive means that default has not happened yet (the present moment being time 0). We admit the possibility that no default will ever take place by including ∞ among the values that τ may attain. However, for bonds with finite maturity $T > 0$, the range of τ can be restricted to $(0, \infty)$ as we are not interested in what happens after the bonds mature.

A random variable can be characterised by its distribution function

$$F_P(t) = P(\tau \leq t).$$

Because the default time τ is positive, we have $F_P(0) = 0$. It is convenient to introduce the probability of survival up to time t,

$$G_P(t) = 1 - F_P(t) = P(t < \tau).$$

An important special case is when τ has a density $f_P(t)$. Since $\tau > 0$ with certainty, we have $f_P(s) = 0$ for $s < 0$. In this case

$$F_P(t) = \int_0^t f_P(s)ds. \tag{2.1}$$

We want to avoid creating a model where default is certain before any given time $t > 0$, which would be unrealistic. Hence we impose the following assumption.

Assumption 2.3
$F_P(t) < 1$ for all $t > 0$.

Example 2.4
Suppose that τ is an exponentially distributed random variable with parameter $\lambda > 0$, which has density

$$f_P(t) = \begin{cases} \lambda e^{-\lambda t} & \text{if } t \geq 0, \\ 0 & \text{otherwise.} \end{cases}$$

The corresponding distribution function is

$$F_P(t) = \begin{cases} 1 - e^{-\lambda t} & \text{if } t \geq 0, \\ 0 & \text{otherwise,} \end{cases}$$

with survival probability

$$G_P(t) = e^{-\lambda t}.$$

The probability that τ is finite is one,

$$P(\tau < \infty) = P\left(\bigcup_{n=1}^{\infty} \{\tau \le n\}\right)$$
$$= \lim_{n \to \infty} P(\tau \le n)$$
$$= \lim_{n \to \infty} F_P(n) = 1.$$

Clearly $F'_p(t) = \lambda e^{-\lambda t}$ for $t > 0$, so $F'_p(t) = f_P(t)$ in each region where F_P is differentiable (i.e. except for $t = 0$, where $f_P(t)$ has a discontinuity). Note that Assumption 2.3 is satisfied.

We can express the parameter λ as

$$\lambda = \frac{f_P(t)}{1 - F_P(t)}.$$

Moreover, using the survival function, we can write

$$\lambda t = -\ln G_P(t).$$

The last expression is an example of what is known as the hazard function, with λ being the so-called hazard rate; see Definitions 2.10 and 2.12.

A random time with exponential distribution can serve to illustrate some key features of the model. In particular, note that in this case the distribution function is continuous and strictly increasing on $[0, \infty)$.

In general, continuity of the distribution function means that all events of the form $\{\tau = t\}$ have probability zero. If the distribution function is strictly increasing on $[0, \infty)$, then the probability of default occurring between any two times s, t such that $0 \le s < t$ is positive. Although one might argue that no default can happen during weekends or holidays, a new piece of information could emerge at any time, and τ indicates this moment rather than the actual instant of bankruptcy, which would follow once the necessary legal procedures are completed. Motivated by these remarks, we adopt the next assumption.

Assumption 2.5
The distribution function $F_P(t)$ of the default time τ is continuous and strictly increasing on $[0, \infty)$.

Remark 2.6
Note that the property $F_P(t) < 1$ adopted earlier follows from the assumption that F_P is strictly increasing.

As a result of the assumptions made so far, $F_P(t)$ is quite regular, but it is not enough to imply that τ must have a density. However, examples with a continuous and increasing $F_P(t)$ but no density can be quite abstract, and assuming the existence of density is convenient in practice.

Assumption 2.7
The random variable τ has a density $f_P(t)$.

A further simplification is provided by the next proposition, which shows that we can restrict our attention to random variables defined on a very specific probability space.

Proposition 2.8
Given τ defined on an arbitrary probability space $(\Omega, \mathcal{F}, P)$, there exists a random variable $\bar{\tau}$ on $\bar{\Omega} = [0, 1]$ equipped with the σ-field of Borel sets and Lebesgue measure m as the probability measure, such that τ and $\bar{\tau}$ have the same probability distribution.

Proof Since the distribution function F_P of τ is strictly increasing on $[0, \infty)$, it is invertible as a function from $[0, \infty)$ to $[0, 1)$, and we can define a random variable $\bar{\tau}$ on $\bar{\Omega} = [0, 1]$ by

$$\bar{\tau}(\omega) = F_P^{-1}(\omega) \quad \text{for any } \omega \in [0, 1),$$

and put any value at $\omega = 1$, say $\bar{\tau}(1) = \infty$. The distribution functions of τ and $\bar{\tau}$ are indeed the same,

$$P(\bar{\tau} \leq t) = m(\{\omega \in [0, 1) : F_P^{-1}(\omega) \leq t\}) = m([0, F_P(t)]) = F_P(t).$$

$\square$

Exercise 2.1 In general, if F_P is just non-decreasing (rather than strictly increasing) on $[0, \infty)$, we can define $\bar{\tau}(\omega) = \sup\{t \geq 0 : F_P(t) < \omega\}$ for any $\omega \in (0, 1]$, and $\bar{\tau}(0) = 0$. Show that τ and $\bar{\tau}$ have the same probability distribution.

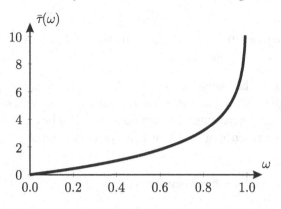

Figure 2.1 The graph of $\bar{\tau}(\omega) = -\frac{1}{\lambda}\ln(1-\omega)$ for $\lambda = 0.5$ in Example 2.9.

Example 2.9
If τ has the exponential distribution, then $F_P(t) = 1 - e^{-\lambda t}$ for $t \geq 0$, so

$$\bar{\tau}(\omega) = -\frac{1}{\lambda}\ln(1-\omega) \quad \text{for any } \omega \in [0, 1)$$

and $\bar{\tau}(1) = \infty$. Figure 2.1 shows the graph of $\bar{\tau}$ for $\lambda = 0.5$.

Exercise 2.2 Show that $\sigma(\bar{\tau})$, the σ-field on $[0, 1]$ generated by $\bar{\tau}$ in Example 2.9, is equal to the family of Borel sets in $[0, 1]$.

As a consequence of Proposition 2.8, we can assume without loss of generality that τ is defined on $\Omega = [0, 1]$ equipped with Lebesgue measure.

Recall that the probability P is the real-life one, hence its properties are related to historical data, which is used in the following exercise. The discussion of survival probabilities only makes sense under real-life probability.

Exercise 2.3 Among 10 similar companies observed over 20 years, two companies defaulted in year 5 and three in year 12. Assume that the default time of each company is determined by an exponentially distributed random variable with common parameter $\lambda > 0$, and these random variables are independent. Estimate λ. What is the probabil-

ity that a given company will survive beyond 5 years? What is the
expected time of default?

Definition 2.10

The **hazard function** is defined as

$$\Gamma_P(t) = -\ln(1 - F_P(t)) = -\ln G_P(t).$$

Recall the assumption that $F_P(t) < 1$ for all t, which guarantees that the
hazard function is a well-defined object. Note that $\Gamma_P(0) = 0$, the function
$\Gamma_P(t)$ is increasing for $t \geq 0$ since $F_P(t)$ is increasing, and $\Gamma_P(t) > 0$ for all
$t > 0$.

Each of the three functions F_P, G_P, Γ_P completely determines the distribution of τ. Given one of these functions, the other two can easily be
computed. The relationship between F_P and G_P is trivial, G_P determines
Γ_P by definition, and just one more relationship is needed to complete the
picture, namely

$$G_P(t) = e^{-\Gamma_P(t)}.$$

The main reason why we need all three functions in what follows is notational convenience. The hazard function is particularly useful as it bears
some resemblance to the discounting exponential, and it will indeed play a
similar role.

The assumptions imposed on F_P immediately imply the following properties.

Proposition 2.11
*The functions $G_P(t)$ and $\Gamma_P(t)$ are continuous, $G_P(t)$ is strictly decreasing,
and $\Gamma_P(t)$ is strictly increasing with $G_P(0) = 1$ and $\Gamma_P(0) = 0$.*

Definition 2.12
If there is a function γ_P such that the hazard function Γ_P can be written as

$$\Gamma_P(t) = \int_0^t \gamma_P(s)ds$$

for each $t \geq 0$, then γ_P is called the **hazard rate**.

Observe that $\gamma_P(t) > 0$ almost everywhere since $\Gamma_P(t)$ is strictly increasing.

Proposition 2.13

If τ has a density f_P, then the hazard rate exists and it is given by

$$\gamma_P(t) = \frac{f_P(t)}{1 - F_P(t)}.$$

Conversely, if the hazard rate γ_P exists, then τ has a density of the form

$$f_P(t) = \gamma_P(t)e^{-\Gamma_P(t)}.$$

Proof Suppose that τ has a density f_P. We want to show that

$$\int_0^t \frac{f_P(s)}{1 - F_P(s)}ds = \Gamma_P(t).$$

To this end we compute the integral on the left,

$$\int_0^t \frac{f_P(s)}{1 - F_P(s)}ds = \int_0^t \frac{F_P'(s)}{1 - F_P(s)}ds = -\ln(1 - F_P(t)) = \Gamma_P(t)$$

since $f_P(s) = F_P'(s)$ for almost every $s \in [0, \infty)$.

Conversely, if the hazard rate γ_P exists, hence $\gamma_P(t) = \Gamma_P'(t)$ for almost every $t \in [0, \infty)$, then we write the distribution function as

$$F_P(t) = 1 - e^{-\Gamma_P(t)},$$

and differentiate with respect to t to get

$$F_P'(t) = \gamma_P(t)e^{-\Gamma_P(t)} = f_P(t)$$

for almost every $t \in [0, \infty)$.   □

Example 2.14

If τ has the exponential distribution, then for all $t \geq 0$

$$\gamma_P(t) = \lambda,$$
$$\Gamma_P(t) = \lambda t.$$

Next, we give an intuitive interpretation of the hazard rate. We compute the probability that a default will take place in the nearest future assuming that it has not happened yet at a given time t. Because of this result, $\gamma_P(t)$ is sometimes called the **instantaneous default rate**.

Proposition 2.15

We have

$$\frac{1}{h}P(\tau \leq t + h | t < \tau) \to \gamma_P(t) \quad as\ h \to 0.$$

Proof   We compute the conditional probability. By definition,

$$P(\tau \le t + h | t < \tau) = \frac{P(\tau \le t + h \text{ and } t < \tau)}{P(t < \tau)}$$

$$= \frac{P(t < \tau \le t + h)}{1 - P(\tau \le t)}$$

$$= \frac{P(\tau \le t + h) - P(\tau \le t)}{1 - P(\tau \le t)},$$

where we use the fact that $\{\tau \le t + h\} \supset \{\tau \le t\}$, so

$$P(\{\tau \le t + h\} \setminus \{\tau \le t\}) = P(\tau \le t + h) - P(\tau \le t).$$

Finally, by the definition of F_P,

$$P(\tau \le t + h | t < \tau) = \frac{F_P(t + h) - F_P(t)}{1 - F_P(t)}.$$

Dividing both sides by h, we have

$$\frac{1}{h} P(\tau \le t + h | t < \tau) \to \frac{F_P'(t)}{1 - F_P(t)} = \gamma_P(t) \quad \text{as } h \to 0.$$

$\square$

Example 2.16

We present a construction of τ based on prescribing a hazard function Γ_P that satisfies the following conditions, which can also be formulated in terms of the distribution function $F_P(t) = 1 - e^{-\Gamma_P(t)}$:

- $\Gamma_P(0) = 0$, so that $F_P(0) = 0$;
- Γ_P is increasing, so that F_P is also increasing;
- Γ_P is continuous, so that the same is true for F_P;
- $\lim_{t \to \infty} \Gamma_P(t) = \infty$, so that $\lim_{t \to \infty} F_P(t) = 1$.

We also need a source of randomness. This will be provided by a random variable θ with uniform distribution in $[0, 1]$. We put

$$\tau = \inf\{t \ge 0 : e^{-\Gamma_P(t)} \le \theta\},$$

that is, we define τ as the first time when the decreasing function $e^{-\Gamma_P(t)}$, which starts at one and goes down to zero, crosses the randomly set barrier θ.

Obviously, $\tau > 0$ almost surely. To see that Γ_P is indeed the hazard function of τ, it is sufficient to consider the survival function

$$G_P(t) = P(t < \tau) = P(e^{-\Gamma_P(t)} > \theta) = P(\theta \in [0, e^{-\Gamma_P(t)})) = e^{-\Gamma_P(t)}.$$

This is an example of the so-called **canonical construction** of the default time τ. In general, the canonical construction starts with a stochastic process Γ_P, though here we have just a particular case with deterministic Γ_P. The general case will be discussed in Section 5.4.

2.3 Default indicator process

In this chapter the only random component of the model is the time of default τ. We will develop a framework to describe the flow of information provided by τ. First we introduce the **default indicator** process

$$I(t) = \mathbf{1}_{\{\tau \le t\}} \quad \text{for } t \ge 0,$$

that is,

$$I(t, \omega) = \begin{cases} 0 & \text{if } t < \tau(\omega) \text{ (before default)}, \\ 1 & \text{if } t \ge \tau(\omega) \text{ (after default)}. \end{cases}$$

This process has non-decreasing right-continuous trajectories, and it jumps from zero to one at the moment of default.

Filtration

The filtration $(\mathcal{I}_t)_{t \ge 0}$ generated by the default indicator process is defined by

$$\mathcal{I}_t = \sigma(\{I(s) : s \in [0, t]\}),$$

that is, $\mathcal{I}_t$ is the smallest σ-field such that all $I(s)$ for $s \le t$ are measurable. Of course, τ is an $(\mathcal{I}_t)_{t \ge 0}$-stopping time since $I(t)$ is $\mathcal{I}_t$-measurable, so

$$\{\tau \le t\} = \{I(t) = 1\} \in \mathcal{I}_t.$$

(Recall that by definition a stopping time satisfies $\{\tau \le t\} \in \mathcal{I}_t$ for each t; see [SCF].)

Exercise 2.4 Prove that $(\mathcal{I}_t)_{t \ge 0}$ is the smallest filtration such that τ is a stopping time with respect to it.

Exercise 2.5 For a fixed $t \geq 0$ list the events belonging to the σ-field $\sigma(I(t))$ generated by default indicator $I(t)$.

We need to understand the structure of the σ-field $\mathcal{I}_t$. Let us start with a specific example before tackling the general setting.

Example 2.17

Consider τ defined on $\Omega = [0, 1]$ by

$$\tau(\omega) = -\frac{1}{\lambda} \ln(1 - \omega) \quad \text{for any } \omega \in [0, 1),$$

and $\tau(1) = \infty$; see Example 2.9.

First observe that for any $s \in [0, t]$

$$[0, 1 - e^{-\lambda s}] = \{\tau \leq s\} = \{I(s) = 1\} \in \mathcal{I}_t,$$
$$(1 - e^{-\lambda s}, 1] = \{s < \tau\} = \{I(s) = 0\} \in \mathcal{I}_t.$$

Now take any interval $(a, b] \subset [0, 1 - e^{-\lambda t}]$ and put $u = \tau(a)$, $v = \tau(b)$. Then

$$(a, b] = (1 - e^{-\lambda u}, 1] \cap [0, 1 - e^{-\lambda v}] \in \mathcal{I}_t$$

since $u, v \leq t$ and so $(1 - e^{-\lambda u}, 1], [0, 1 - e^{-\lambda v}] \in \mathcal{I}_t$.

Because $(a, b] \in \mathcal{I}_t$ for every interval $(a, b] \subset [0, 1 - e^{-\lambda t}]$, it follows that the σ-field $\mathcal{I}_t$ contains every Borel set $B \subset [0, 1 - e^{-\lambda t}]$. Moreover, because $(1 - e^{-\lambda t}, 1] \in \mathcal{I}_t$, it follows that $\mathcal{I}_t$ also contains every set of the form $B \cup (1 - e^{-\lambda t}, 1]$, where B is a Borel set such that $B \subset [0, 1 - e^{-\lambda t}]$. It turns out that $\mathcal{I}_t$ contains no other sets. This will be shown in detail in the proof of Proposition 2.19.

We observe that the set $(1 - e^{-\lambda t}, 1] = \{t < \tau\}$ is an **atom** in $\mathcal{I}_t$, which by definition means that the only subsets of $\{t < \tau\}$ that belong to $\mathcal{I}_t$ are $\emptyset$ and $\{t < \tau\}$ itself. Indeed, any set in $\mathcal{I}_t$ must be of the form B or $B \cup (1 - e^{-\lambda t}, 1]$ for some Borel set $B \subset [0, 1 - e^{-\lambda t}]$. But since the intervals $[0, 1 - e^{-\lambda t}]$ and $(1 - e^{-\lambda t}, 1]$ are disjoint, we must have $B = \emptyset$ if B or $B \cup (1 - e^{-\lambda t}, 1]$ is a subset of $(1 - e^{-\lambda t}, 1] = \{t < \tau\}$.

Finally, for every Borel set $B \subset [0, 1 - e^{-\lambda t}]$ we can write

$$B = C \cap [0, 1 - e^{-\lambda t}] = C \cap \{\tau \leq t\},$$
$$B \cup (1 - e^{-\lambda t}, 1] = D \cup (1 - e^{-\lambda t}, 1] = D \cup \{t < \tau\}$$

for some Borel sets $C, D \subset [0, 1]$. This enables us to express $\mathcal{I}_t$ as

$$\mathcal{I}_t = \{C \cap \{\tau \leq t\} : C \subset [0, 1] \text{ is a Borel set}\}$$
$$\cup \{D \cup \{t < \tau\} : D \subset [0, 1] \text{ is a Borel set}\}.$$

In Example 2.17 $\sigma(\tau)$ happens to coincide with the family of Borel sets in $[0, 1]$, that is, with the σ-field on the underlying probability space $\Omega = [0, 1]$. This does not have to hold in general; $\sigma(\tau)$ can in fact be smaller than the σ-field of the probability space. This will be so in the next example, and we shall see that in such a case the range of the sets C, D should be $\sigma(\tau)$.

Example 2.18

To provide a simple example of a situation where $\sigma(\tau)$ differs from the σ-field of the probability space, consider $\Omega = [0, 1] \times [0, 1]$ with the σ-field of Borel subsets of $[0, 1] \times [0, 1]$ and two-dimensional Lebesgue measure. Let

$$\tau(\omega_1, \omega_2) = -\frac{1}{\lambda} \ln(1 - \omega_1) \quad \text{for any } (\omega_1, \omega_2) \in [0, 1) \times [0, 1],$$

and $\tau(1, \omega_2) = \infty$ for any $\omega_2 \in [0, 1]$. Now

$$\{I(s) = 1\} = \{\tau \leq s\} = [0, 1 - e^{-\lambda t}] \times [0, 1] \in \mathcal{I}_t,$$
$$\{I(s) = 0\} = \{s < \tau\} = (1 - e^{-\lambda t}, 1] \times [0, 1] \in \mathcal{I}_t,$$

and consequently

$$\mathcal{I}_t = \{(C \times [0, 1]) \cap \{\tau \leq t\} : C \subset [0, 1] \text{ is a Borel set}\}$$
$$\cup \{(D \times [0, 1]) \cup \{t < \tau\} : D \subset [0, 1] \text{ is a Borel set}\}.$$

Since

$$\sigma(\tau) = \{B \times [0, 1] : B \subset [0, 1] \text{ is a Borel set}\},$$

we can write $\mathcal{I}_t$ as

$$\mathcal{I}_t = \{C \cap \{\tau \leq t\} : C \in \sigma(\tau)\} \cup \{D \cup \{t < \tau\} : D \in \sigma(\tau)\},$$

but not as

$$\{C \cap \{\tau \leq t\} : C \subset [0, 1] \times [0, 1] \text{ is a Borel set}\}$$
$$\cup \{D \cup \{t < \tau\} : D \subset [0, 1] \times [0, 1] \text{ is a Borel set}\}$$

because $\sigma(\tau)$ differs from the family of Borel sets on $[0, 1] \times [0, 1]$.

In general, we have the following description of the σ-field $\mathcal{I}_t$.

Proposition 2.19

Let $t \geq 0$. Then

(i) *$\mathcal{I}_t$ consists of all sets of the form $A \cap \{\tau \leq t\}$ for some $A \in \sigma(\tau)$ and their complements, which are of the form $B \cup \{t < \tau\}$ for some $B \in \sigma(\tau)$, and only of such sets;*

(ii) *$\{t < \tau\}$ is an atom of $\mathcal{I}_t$.*

Proof (i) Take a set of the form $A \cap \{\tau \leq t\}$ with $A \in \sigma(\tau)$ and compute its complement. By de Morgan's law,

$$\Omega \setminus (A \cap \{\tau \leq t\}) = (\Omega \setminus A) \cup \{t < \tau\},$$

so this complement is of the form $B \cup \{t < \tau\}$ for some $B \in \sigma(\tau)$ (namely $B = \Omega \setminus A \in \sigma(\tau)$). We introduce two families of sets

$$\mathcal{D}_1 = \{A \cap \{\tau \leq t\} : A \in \sigma(\tau)\},$$
$$\mathcal{D}_2 = \{B \cup \{t < \tau\} : B \in \sigma(\tau)\}.$$

As we know, $\mathcal{D}_2$ consists of the complements of sets in $\mathcal{D}_1$. Vice versa, take an element $B \cup \{t < \tau\}$ of $\mathcal{D}_2$, with $B \in \sigma(\tau)$, and consider its complement

$$\Omega \setminus (B \cup \{t < \tau\}) = (\Omega \setminus B) \cap \{\tau \leq t\}$$

to see that it belongs to $\mathcal{D}_1$. (In particular, this implies that the families $\mathcal{D}_1$ and $\mathcal{D}_2$ are disjoint.)

Thus, assertion (i) reads

$$\mathcal{I}_t = \mathcal{D}_1 \cup \mathcal{D}_2.$$

To prove this, first we show that the union $\mathcal{D}_1 \cup \mathcal{D}_2$ is a σ-field. To see that $\emptyset \in \mathcal{D}_1$ take $A = \emptyset$, and to see that $\Omega \in \mathcal{D}_2$ take $B = \Omega$. Next, $\mathcal{D}_1 \cup \mathcal{D}_2$ is closed under complements (as noted above). Moreover, it a routine exercise (see Exercise 2.6) to show that $\mathcal{D}_1$ is closed under countable unions of sets, and so is $\mathcal{D}_2$. In addition, we shall see that if $C \in \mathcal{D}_1$ and $D \in \mathcal{D}_2$, then $C \cup D \in \mathcal{D}_2$. To this end let $C = A \cap \{\tau \leq t\}$, $D = B \cup \{t < \tau\}$ for some $A, B \in \sigma(\tau)$, and note that

$$\begin{aligned} C \cup D &= (A \cap \{\tau \leq t\}) \cup (B \cup \{t < \tau\}) \\ &= (A \cup B \cup \{t < \tau\}) \cap (\{\tau \leq t\} \cup B \cup \{t < \tau\}) \\ &= A \cup B \cup \{t < \tau\} \in \mathcal{D}_2. \end{aligned}$$

This implies that $\mathcal{D}_1 \cup \mathcal{D}_2$ is also closed under countable unions, and this shows that $\mathcal{D}_1 \cup \mathcal{D}_2$ is a σ-field.

Next, observe that $I(s)$ is measurable with respect to the σ-field $\mathcal{D}_1 \cup \mathcal{D}_2$ for any $s \leq t$. This follows from the fact that $\mathcal{D}_1$ contains $\{\tau \leq s\}$ for each $s \leq t$ (take $A = \{\tau \leq s\}$).

These two facts imply that

$$\mathcal{I}_t \subset \mathcal{D}_1 \cup \mathcal{D}_2$$

since $\mathcal{I}_t$ is by definition the smallest σ-field with respect to which $I(s)$ is measurable for each $s \leq t$.

To verify the reverse inclusion

$$\mathcal{I}_t \supset \mathcal{D}_1 \cup \mathcal{D}_2$$

we will show that $\mathcal{D}_1 \subset \mathcal{I}_t$ and $\mathcal{D}_2 \subset \mathcal{I}_t$.

For the first inclusion consider an element of $\sigma(\tau)$ of the form $A = \{\tau \leq u\}$ for any $u \in \mathbb{R}$. The corresponding element of $\mathcal{D}_1$ has the form

$$A \cap \{\tau \leq t\} = \{\tau \leq u\} \cap \{\tau \leq t\}.$$

If $u \leq t$, then

$$A \cap \{\tau \leq t\} = \{\tau \leq u\} \in \sigma(I(u)) \subset \mathcal{I}_t,$$

and, if $u > t$, then

$$A \cap \{\tau \leq t\} = \{\tau \leq t\} \in \sigma(I(t)) \subset \mathcal{I}_t.$$

We have proved that the generators of $\sigma(\tau)$ produce elements of $\mathcal{I}_t$. To see that this is also true for any element of $\sigma(\tau)$ it is sufficient to see that the family

$$\mathcal{D} = \{C \in \sigma(\tau) : C \cap \{\tau \leq t\} \in \mathcal{I}_t\} \subset \sigma(\tau)$$

is a σ-field, since this shows that $\sigma(\tau) \subset \mathcal{D}$ ($\sigma(\tau)$ is the smallest σ-field containing the generators) so $\sigma(\tau) = \mathcal{D}$, which implies that all events of the form $C \cap \{\tau \leq t\}$ for some $C \in \sigma(\tau)$ are in $\mathcal{I}_t$, that is, $\mathcal{D}_1 \subset \mathcal{I}_t$.

To see that $\mathcal{D}$ is a σ-field, note that $\varnothing \cap \{\tau \leq t\} = \varnothing \in \mathcal{I}_t$ and $\Omega \cap \{\tau \leq t\} = \{\tau \leq t\} \in \mathcal{I}_t$, so $\varnothing, \Omega \in \mathcal{D}$. Furthermore, if $C \in \mathcal{D}$, then

$$(\Omega \setminus C) \cap \{\tau \leq t\} = \{\tau \leq t\} \setminus (C \cap \{\tau \leq t\}) \in \mathcal{I}_t,$$

so $\Omega \setminus C \in \mathcal{D}$. Verification of the condition for countable unions is routine; see Exercise 2.7.

It remains to show that $\mathcal{D}_2 \subset \mathcal{I}_t$. For each $B \in \sigma(\tau)$

$$B \cup \{t < \tau\} = (B \cap \{\tau \leq t\}) \cup \{t < \tau\} \in \mathcal{I}_t$$

since $B \cap \{\tau \leq t\} \in \mathcal{D}_1 \subset \mathcal{I}_t$ and $\{t < \tau\} \in \sigma(I(t)) \subset \mathcal{I}_t$. So $\mathcal{D}_2 \subset \mathcal{I}_t$, and this concludes the proof of the first claim.

(ii) We have to show that $\{t < \tau\}$ has only trivial subsets within $\mathcal{I}_t$, that is, $\varnothing$ and $\{t < \tau\}$ are the only elements of $\mathcal{I}_t$ among the subsets of $\{t < \tau\}$. To this end, take any $C \in \mathcal{I}_t$ and suppose that $C \subset \{t < \tau\}$. We have to show that either $C = \varnothing$ or $C = \{t < \tau\}$. Since $\mathcal{I}_t = \mathcal{D}_1 \cup \mathcal{D}_2$, either $C = A \cap \{\tau \leq t\}$ for some $A \in \sigma(\tau)$, or $C = B \cup \{t < \tau\}$ for some $B \in \sigma(\tau)$. In the first case, to have $A \cap \{\tau \leq t\} \subset \{t < \tau\}$ the only possibility is that $A = \varnothing$, thus $C = \varnothing$. In the other case, to have $B \cup \{t < \tau\} \subset \{t < \tau\}$ we must have $B \subset \{t < \tau\}$, hence $C = \{t < \tau\}$. □

Exercise 2.6   Show that $\mathcal{D}_1$ and $\mathcal{D}_2$ (defined in the proof of Proposition 2.19) are closed under countable unions of sets.

Exercise 2.7   Show that $\mathcal{D}$ (defined in the proof of Proposition 2.19) is closed under countable unions of sets.

Corollary 2.20
For all $t \geq 0$ we have $\mathcal{I}_t \subset \sigma(\tau)$.

Proof   For any $A \in \sigma(\tau)$ we have $A \cap \{\tau \leq t\} \in \sigma(\tau)$ since $\{\tau \leq t\} \in \sigma(\tau)$. This implies that $\mathcal{D}_1 \subset \sigma(\tau)$ (the family $\mathcal{D}_1$ is defined in the proof of Proposition 2.19). The complement of $A \cap \{\tau \leq t\}$ also belongs to $\sigma(\tau)$ since $\sigma(\tau)$ is a σ-field. The claim follows from Proposition 2.19. □

Remark 2.21
Corollary 2.20 has some important consequences. In general, the payoff of a European derivative security with exercise time T is a random variable measurable with respect to the corresponding σ-field in the filtration generated by the underlying process. If the underlying process is taken to be the default indicator $I(t)$, any such payoff will be $\mathcal{I}_T$-measurable and so $\sigma(\tau)$-measurable, hence of the form $h(\tau)$ for some Borel function h (see [PF]). At this moment we assume the underlying process to be $I(t)$; however, we would prefer to use $D(t, T)$ in this role. As we shall see in Remark 2.41, this process of prices generates the same filtration as $I(t)$.

Exercise 2.8   Show that $I_t = \sigma(\{t < \tau\}, \sigma(\tau \wedge t))$.

Exercise 2.9   Let $I_\infty = \sigma(\bigcup_{t \geq 0} I_t)$. Show that $I_\infty = \sigma(\tau)$.

Exercise 2.10   Show that τ is not I_t-measurable (i.e. $\sigma(\tau) \not\subset I_t$) for any t.

Conditional expectation

Due to the structure of I_t, an I_t-measurable random variable must have a very special form.

Proposition 2.22
If Z is I_t-measurable, then

$$Z = \eta_t(\tau)\mathbf{1}_{\{\tau \leq t\}} + c_t \mathbf{1}_{\{t < \tau\}}$$

for some Borel function $\eta_t : \mathbb{R} \to \mathbb{R}$ and a deterministic number c_t.

Proof   Since $I_t \subset \sigma(\tau)$ by Corollary 2.20, if Z is I_t-measurable, then it is also $\sigma(\tau)$-measurable. As a result, we know (see [PF]) that there is a Borel function $\eta_t : \mathbb{R} \to \mathbb{R}$ such that

$$Z = \eta_t(\tau). \tag{2.2}$$

According to assertion (ii) of Proposition 2.19, $\{t < \tau\}$ is an atom in I_t. This means that the random variable Z, which is I_t-measurable, must be constant on $\{t < \tau\}$. We denote this constant value by c_t. Together with (2.2) this concludes the proof.   □

Remark 2.23
We can merge η_t and c_t into a single Borel function $\bar{\eta}_t : \mathbb{R} \to \mathbb{R}$ defined as $\bar{\eta}_t(s) = \eta(s)$ for $s \leq t$ and $\bar{\eta}_t(s) = c_t$ for $s > t$. Then

$$Z = \bar{\eta}_t(\tau)\mathbf{1}_{\{\tau \leq t\}} + \bar{\eta}_t(\tau)\mathbf{1}_{\{t < \tau\}} = \bar{\eta}_t(\tau).$$

Observe that $\bar{\eta}_t$ is constant on (t, ∞).

The next proposition gives a useful formula for conditional expectation. It will be applied to find the value at time t of a future random payoff X to be paid only if no default happens.

Proposition 2.24

If X is an integrable random variable, then for all $t \geq 0$

$$\mathbb{E}_P(X\mathbf{1}_{\{t<\tau\}}|\mathcal{I}_t) = \mathbf{1}_{\{t<\tau\}}\frac{\mathbb{E}_P(X\mathbf{1}_{\{t<\tau\}})}{P(t<\tau)} = \mathbf{1}_{\{t<\tau\}}\mathbb{E}_P(X|t<\tau). \qquad (2.3)$$

Proof Clearly, $\mathbb{E}_P(X\mathbf{1}_{\{t<\tau\}}|\mathcal{I}_t)$ is $\mathcal{I}_t$-measurable, so we can use Proposition 2.22 to write

$$\mathbb{E}_P(X\mathbf{1}_{\{t<\tau\}}|\mathcal{I}_t) = \eta_t(\tau)\mathbf{1}_{\{\tau\leq t\}} + c_t\mathbf{1}_{\{t<\tau\}}$$

for some Borel function η_t and a constant c_t. Since $\{t<\tau\} \in \mathcal{I}_t$,

$$\mathbb{E}_P(X\mathbf{1}_{\{t<\tau\}}|\mathcal{I}_t) = \mathbf{1}_{\{t<\tau\}}\mathbb{E}_P(X|\mathcal{I}_t),$$

which implies that $\mathbb{E}_P(X\mathbf{1}_{\{t<\tau\}}|\mathcal{I}_t)$ is zero outside the event $\{t<\tau\}$, so

$$\mathbb{E}_P(X\mathbf{1}_{\{t<\tau\}}|\mathcal{I}_t) = c_t\mathbf{1}_{\{t<\tau\}}.$$

Taking the expectation, we get

$$\mathbb{E}_P(X\mathbf{1}_{\{t<\tau\}}) = c_t\mathbb{E}_P(\mathbf{1}_{\{t<\tau\}}) = c_tP(t<\tau),$$

which shows that

$$c_t = \frac{\mathbb{E}_P(X\mathbf{1}_{\{t<\tau\}})}{P(t<\tau)} = \mathbb{E}_P(X|t<\tau).$$

This completes the proof. □

Remark 2.25

We have the following alternative forms of the conditional expectation in Proposition 2.24:

$$\mathbb{E}_P(X\mathbf{1}_{\{t<\tau\}}|\mathcal{I}_t) = \mathbf{1}_{\{t<\tau\}}\frac{\mathbb{E}_P(X\mathbf{1}_{\{t<\tau\}})}{G_P(t)} = \mathbf{1}_{\{t<\tau\}}e^{\Gamma_P(t)}\mathbb{E}_P(X\mathbf{1}_{\{t<\tau\}}).$$

If we know the distribution function of the default time, Proposition 2.24 gives an expression for the probability of survival beyond time t given the information available at an earlier time s.

Corollary 2.26

If $s < t$, then

$$\mathbb{E}_P(\mathbf{1}_{\{t<\tau\}}|\mathcal{I}_s) = \mathbf{1}_{\{s<\tau\}}\frac{1-F_P(t)}{1-F_P(s)} = \mathbf{1}_{\{s<\tau\}}\frac{G_P(t)}{G_P(s)} = \mathbf{1}_{\{s<\tau\}}e^{-(\Gamma_P(t)-\Gamma_P(s))}.$$

Proof Take $X = \mathbf{1}_{\{t < \tau\}}$. The left-hand side of (2.3) becomes

$$\mathbb{E}_P(\mathbf{1}_{\{t<\tau\}}\mathbf{1}_{\{s<\tau\}}|\mathcal{I}_s) = \mathbb{E}_P(\mathbf{1}_{\{t<\tau\}}|\mathcal{I}_s)$$

since $\{t < \tau\} \subset \{s < \tau\}$. On the other hand, the right-hand side of (2.3) becomes

$$\mathbf{1}_{\{s<\tau\}}\frac{\mathbb{E}_P(\mathbf{1}_{\{t<\tau\}}\mathbf{1}_{\{s<\tau\}})}{P(s<\tau)} = \mathbf{1}_{\{s<\tau\}}\frac{\mathbb{E}_P(\mathbf{1}_{\{t<\tau\}})}{P(s<\tau)} = \mathbf{1}_{\{s<\tau\}}\frac{1 - F_P(t)}{1 - F_P(s)},$$

which proves the first equality. The remaining equalities are obvious reformulations.   □

Proposition 2.24 solves the problem of finding the conditional expectation, but only on the event $\{t < \tau\}$. We give the full formula below.

Proposition 2.27
If X is an integrable random variable, then

$$\mathbb{E}_P(X|\mathcal{I}_t) = \mathbf{1}_{\{\tau\leq t\}}\mathbb{E}_P(X|\sigma(\tau)) + \mathbf{1}_{\{t<\tau\}}\frac{\mathbb{E}_P(X\mathbf{1}_{\{t<\tau\}})}{P(t<\tau)}. \qquad (2.4)$$

Proof We can write $X = X\mathbf{1}_{\{\tau\leq t\}} + X\mathbf{1}_{\{t<\tau\}}$. Proposition 2.24 gives the conditional expectation of $X\mathbf{1}_{\{t<\tau\}}$. It remains to show that

$$\mathbb{E}_P(X\mathbf{1}_{\{\tau\leq t\}}|\mathcal{I}_t) = \mathbb{E}_P(X\mathbf{1}_{\{\tau\leq t\}}|\sigma(\tau)).$$

If $A \in \sigma(\tau)$, then $A \cap \{\tau \leq t\} \in \mathcal{I}_t$ by Proposition 2.19. Applying the definition of conditional expectation, we get

$$\int_A \mathbb{E}_P(X\mathbf{1}_{\{\tau\leq t\}}|\sigma(\tau))dP = \int_A X\mathbf{1}_{\{\tau\leq t\}}dP = \int_{A\cap\{\tau\leq t\}} XdP = \int_{A\cap\{\tau\leq t\}} \mathbb{E}_P(X|\mathcal{I}_t)dP$$

$$= \int_A \mathbf{1}_{\{\tau\leq t\}}\mathbb{E}_P(X|\mathcal{I}_t)dP = \int_A \mathbb{E}_P(X\mathbf{1}_{\{\tau\leq t\}}|\mathcal{I}_t)dP,$$

which completes the argument (note that $\mathbf{1}_{\{\tau\leq t\}}$ is both $\mathcal{I}_t$-measurable and $\sigma(\tau)$-measurable).   □

For random variables of some special form which are of interest for us, we can obtain more explicit expressions in terms of the functions G_P, Γ_P, or F_P characterising τ.

Proposition 2.28
Suppose that $X = h(\tau)$ for a Borel function h. Then

$$\mathbb{E}_P(X|\mathcal{I}_t) = \mathbf{1}_{\{\tau\leq t\}}h(\tau) + \mathbf{1}_{\{t<\tau\}}\int_t^\infty h(u)e^{\Gamma_P(t)-\Gamma_P(u)}\gamma_P(u)du.$$

Proof We apply Proposition 2.27. The form of the first term on the right-hand side follows from the measurability of $h(\tau)$ with respect to $\sigma(\tau)$. For the second term we just note that

$$\mathbb{E}_P(X\mathbf{1}_{\{t<\tau\}}) = \int_t^\infty h(u)f_P(u)du = \int_t^\infty h(u)\gamma_P(u)e^{-\Gamma_P(u)}du$$

due to the relation $f_P(t) = \gamma_P(t)e^{-\Gamma_P(t)}$ between f_P and γ_P established in Proposition 2.13. □

2.4 No arbitrage and risk-neutral probability

Recall that the price of a defaultable zero-coupon unit bond is denoted by $D(t,T)$ for $t \le T$, where t is the running time and T is the maturity date of the bond. If default occurs before or at maturity, that is, if $\tau \le T$, then we assume that there is no compensation, the bond holder receives nothing. The only source of randomness is τ, so we also make a natural measurability assumption for $D(t,T)$, with a mild regularity condition added.

Assumption 2.29
We assume that $D(t,T)$ is a stochastic process adapted to the filtration $(\mathcal{I}_t)_{t\ge0}$, with right-continuous paths, and

$$D(T,T) = \mathbf{1}_{\{T<\tau\}}.$$

Our first goal is to discover some properties of the process $D(t,T)$ related to the No Arbitrage Principle. In addition, in line with the classical theories, we would like to find a probability measure Q equivalent to P such that the discounted bond prices $e^{-rt}D(t,T)$ would follow a martingale under Q, so the bond price can be computed as the expectation of the discounted payoff under Q, that is, $D(0,T) = e^{-rT}\mathbb{E}_Q(\mathbf{1}_{\{T<\tau\}})$.

In practice we want, in fact, to do the opposite. Given the bond prices $D(0,T)$, we want to find the characteristics of the risk-neutral probability Q to calibrate the model. Such a probability measure would serve as a pricing tool to value other more complex defaultable securities.

No Arbitrage Principle

The gist of the notion of arbitrage is that, having started with nothing, a positive payoff can be obtained with positive probability and without any possibility of a loss.

To formalise, we have to introduce trading strategies. We write $\varphi_B(t)$ for the number of default-free bonds held at time t, with the same maturity as the defaultable bond. We also write $\varphi_D(t)$ for the number of defaultable bonds held at time t. The time t value of a portfolio $\varphi(t) = (\varphi_B(t), \varphi_D(t))$ of non-defaultable and defaultable bonds is

$$V_\varphi(t) = \varphi_B(t)B(t,T) + \varphi_D(t)D(t,T).$$

Remark 2.30

The classical approach uses the position in units of the money market account $A(t) = e^{rt}$ rather than zero-coupon bonds. Since the interest rate is constant, these two approaches are equivalent. The money market account and $B(t,T)$ grow exponentially at the same rate r, the only difference being the initial value at time 0, which is 1 for the money market account or e^{-rT} for the default-free bond.

Both $\varphi_B(t)$ and $\varphi_D(t)$ are allowed to be stochastic processes. We introduce a condition in line with the classical models, where we assume that any investment decisions at time t are based on the information available at that time. With the random time τ being the only source of uncertainty, this can be formulated as follows.

Assumption 2.31

The processes $\varphi_B(t)$ and $\varphi_D(t)$ are adapted to the filtration $(\mathcal{I}_t)_{t\geq 0}$.

Next, we want to eliminate the possibility of injecting or withdrawing money from the strategy. This is captured by the self-financing condition. However, at this stage we cannot allow rebalancing of the portfolio continuously in time because we have not described the dynamics of $D(t,T)$ yet. We need to restrict the class of strategies to those with rebalancing at a finite number of times.

Definition 2.32

A strategy is called **simple** if there exists a sequence of times $S_k \in [0, T]$ for $k = 0, 1, \ldots, n$ such that $0 = S_0 < S_1 < \cdots < S_n = T$ and
  (i) the processes $\varphi_B(t), \varphi_D(t)$ are constant on $(S_{k-1}, S_k]$ for each $k = 1, \ldots, n$;
  (ii) the random variables $\varphi_B(S_k), \varphi_D(S_k)$ are $\mathcal{I}_{S_{k-1}}$-measurable for each $k = 1, \ldots, n$, and $\varphi_B(0), \varphi_D(0)$ are $\mathcal{I}_0$-measurable (i.e. deterministic).

The self-financing condition can now be formulated for simple strategies of this kind. It is definitely true that no funds are injected or withdrawn within the periods during which the positions in the strategy are constant, so all that is needed is to take care of the times when the strategy can be rebalanced.

Definition 2.33

A **self-financing** simple strategy $\varphi(t) = (\varphi_B(t), \varphi_D(t))$ satisfies the condition

$$V_\varphi(S_k) = \varphi_B(S_{k+1})B(S_k, T) + \varphi_D(S_{k+1})D(S_k, T) \tag{2.5}$$

for each $k = 0, \ldots, n-1$.

This is in fact the same condition as in discrete-time models.

Note that $\varphi_B(t)$ and $\varphi_D(t)$ are left-continuous processes, whereas the value process $V_\varphi(t)$ of a self-financing simple strategy φ is right-continuous.

Definition 2.34

A **simple arbitrage strategy** is a self-financing simple strategy $\varphi(t) = (\varphi_B(t), \varphi_D(t))$ such that $V_\varphi(0) = 0$, $V_\varphi(T) \geq 0$, and $V_\varphi(T) > 0$ with positive probability.

The following assumption, called the **No Simple Arbitrage Principle**, is justified by the requirement to have an economically viable model.

Assumption 2.35

There are no simple arbitrage strategies.

Remark 2.36

The No Simple Arbitrage Principle will be used to prove various facts. The proofs are typically by *reductio ad absurdum*, where the argument is based on finding an arbitrage strategy to arrive at a contradiction. If the notion of arbitrage is restrictive, that is, the class of arbitrage strategies is narrow (as in the case of simple no-arbitrage strategies), this is a relatively more difficult task as compared with the case of a larger class of strategies, and therefore the argument will remain valid when the class of arbitrage strategies is enlarged.

We shall see that the No Simple Arbitrage Principle together with adaptedness can provide quite a detailed description of $D(t, T)$.

Proposition 2.37

For each $t \in [0, T]$ the random variable $D(t, T)$ is almost surely zero on the event $\{\tau \leq t\}$. On the event $\{t < \tau\}$ the price $D(t, T)$ is deterministic and positive.

Proof For $t = T$ we have $D(T, T) = \mathbf{1}_{\{T < \tau\}}$, and the assertion of the proposition is obvious. In the remainder of the proof we take $t < T$.

First of all, we show that $D(t, T) \geq 0$ almost surely. If the event $A = \{D(t, T) < 0\}$ had positive probability, we could achieve arbitrage by taking a long position in the defaultable bond and investing the proceeds in the non-defaultable bond. To be precise, we could take

$$\varphi_D(s) = 0, \qquad \varphi_B(s) = 0 \qquad\qquad \text{for each } s \in [0, t],$$

$$\varphi_D(s) = \mathbf{1}_A, \qquad \varphi_B(s) = -\mathbf{1}_A \frac{D(t, T)}{B(t, T)} \qquad \text{for each } s \in (t, T].$$

This is a self-financing simple strategy with initial value $V_\varphi(0) = 0$ and final value

$$V_\varphi(T) = \mathbf{1}_A \left(\mathbf{1}_{\{T < \tau\}} - \frac{D(t, T)}{B(t, T)}\right),$$

which is non-negative and strictly positive on A, contradicting the No Simple Arbitrage Principle.

Having established that $D(t, T)$ must be non-negative, we observe that the defaultable bond will pay nothing at maturity on the event $\{\tau \leq t\}$, which is contained in $\{\tau \leq T\}$. This intuitively implies that the defaultable bond must be worthless at time t on the event $\{\tau \leq t\}$. Otherwise, arbitrage could be achieved by shorting the defaultable bond if its price were positive, and investing the proceeds in the non-defaultable bond. There would be no need to close the short position at maturity, and the investment in the non-defaultable bond would produce an arbitrage profit. Formally, if the event $B = \{\mathbf{1}_{\{\tau \leq t\}} D(t, T) > 0\}$ had positive probability, we could construct the self-financing simple strategy

$$\varphi_D(s) = 0, \qquad \varphi_B(s) = 0 \qquad\qquad \text{for each } s \in [0, t],$$

$$\varphi_D(s) = -\mathbf{1}_B, \qquad \varphi_B(s) = \mathbf{1}_B \frac{D(t, T)}{B(t, T)} \qquad \text{for each } s \in (t, T],$$

with initial value $V_\varphi(0) = 0$ and final value

$$V_\varphi(T) = \mathbf{1}_B \frac{D(t, T)}{B(t, T)},$$

which is non-negative and strictly positive on B, once again contradicting the No Simple Arbitrage Principle. We conclude that $D(t, T) = 0$ almost surely on $\{\tau \leq t\}$.

Finally, since $D(t, T)$ is $\mathcal{I}_t$-measurable, by Proposition 2.22 there is a deterministic constant c_t such that $D(t, T) = c_t$ on $\{t < \tau\}$. We already know that $c_t \geq 0$ since $D(t, T) \geq 0$ almost surely, and we need to show that, in fact, $c_t > 0$. Suppose that $c_t = 0$. In that case we could construct the simple strategy

$$\varphi_D(s) = 0, \qquad \varphi_B(s) = 0 \qquad \text{for each } s \in [0, t],$$
$$\varphi_D(s) = \mathbf{1}_{\{t < \tau\}}, \qquad \varphi_B(s) = 0 \qquad \text{for each } s \in (t, T].$$

It is a self-financing strategy if $c_t = 0$. It has initial value $V_\varphi(0) = 0$ and terminal value

$$V_\varphi(T) = \mathbf{1}_{\{T < \tau\}},$$

which is non-negative and strictly positive on the event $\{T < \tau\}$, whose probability is positive, leading to yet another contradiction of the No Simple Arbitrage Principle. This completes the proof. $\qquad\square$

In practice defaultable bonds produce a higher yield than the risk-free rate. This means that the defaultable bond is less expensive than the non-defaultable one. This fact is also a consequence of the No Simple Arbitrage Principle.

Proposition 2.38
For all $t \in [0, T)$ we have $D(t, T) < B(t, T)$.

Proof Fix $t \in [0, T)$ and suppose that $A = \{D(t, T) \geq B(t, T)\}$ has positive probability. The idea of constructing arbitrage is straightforward: buy cheap and sell expensive once the opportunity becomes available. So we take

$$\varphi_D(s) = 0, \qquad \varphi_B(s) = 0 \qquad \text{for each } s \in [0, t],$$
$$\varphi_D(s) = -\mathbf{1}_A \frac{B(t, T)}{D(t, T)}, \qquad \varphi_B(s) = \mathbf{1}_A \qquad \text{for each } s \in (t, T].$$

This is a self-financing simple strategy. The initial value is $V_\varphi(0) = 0$ and the terminal value is non-negative,

$$V_\varphi(T) = \mathbf{1}_A \left(1 - \frac{B(t, T)}{D(t, T)} D(T, T) \right) \geq 0$$

since $D(t, T) \geq B(t, T)$ on A.

It remains to show that $V_\varphi(T)$ is strictly positive with positive probability. If $D(T, T) = 0$, then

$$V_\varphi(T) = 1$$

on A, so this value will be attained on $A \cap \{\tau \leq T\}$. Since $D(t, T) = 0$ almost surely on $\{\tau \leq t\}$ while $D(t, T) > 0$ on A, it follows that $A \subset \{t < \tau\}$. But

$A \in \mathcal{I}_t$, and $\{t < \tau\}$ is an atom of $\mathcal{I}_t$, so $A = \{t < \tau\}$ since A has positive probability, hence it is non-empty. Consequently,

$$A \cap \{\tau \le T\} = \{t < \tau \le T\}.$$

The probability of this event equals $F_P(T) - F_P(t)$, a positive number since F is assumed to be strictly increasing. This contradicts the No Simple Arbitrage Principle.   □

Because $B(t, T) = e^{-r(T-t)}$ and $D(t, T)$ is a positive deterministic number such that $D(t, T) < B(t, T)$ on $\{t < \tau\}$ for any $t < T$, we can write

$$\mathbf{1}_{\{t<\tau\}} D(t, T) = \mathbf{1}_{\{t<\tau\}} e^{-r(T-t)-g(t)} \qquad (2.6)$$

for some deterministic function $g : [0, T] \to \mathbb{R}$ such that $g(t) > 0$ if $t \le T$. Moreover, $g(T) = 0$ since $D(T, T) = B(T, T) = 1$ on $\{T < \tau\}$.

Consider two time instants $s, t \in [0, T]$ such that $s < t$ and compare the corresponding bond prices on the event $\{t < \tau\}$ by analysing their ratio:

$$\mathbf{1}_{\{t<\tau\}} \frac{D(t, T)}{D(s, T)} = \mathbf{1}_{\{t<\tau\}} e^{-r(T-t)-g(t)} e^{r(T-s)+g(s)}$$
$$= \mathbf{1}_{\{t<\tau\}} e^{r(t-s)+(g(s)-g(t))}.$$

The difference $g(s) - g(t)$ is responsible for the change of defaultable bond prices as an additional factor to the growth at the risk-free rate. If this difference is positive, the defaultable bond grows faster; otherwise, the default-free bond performs better. The latter case is economically flawed as the risk of default should be compensated by faster growth. This intuition can be formalised.

Proposition 2.39
The function g is strictly decreasing: if $s, t \in [0, T]$ and $s < t$, then $g(s) > g(t)$.

Proof Suppose that $g(s) \le g(t)$ for some $s, t \in [0, T]$ such that $s < t$. This means that between the times s and t the default-free bond grows at least as fast as the defaultable one. We should therefore short the defaultable bond and invest in the risk-free bond between s and t. Then, at time t we move all funds into non-defaultable bonds. If a default occurs between s and t, then we make a profit since we do not need to close the short position. Otherwise, we at least break even. To be more specific, we construct the strategy

$$\varphi_D(u) = 0, \quad \varphi_B(u) = 0 \qquad\qquad\qquad\text{for each } u \in [0, s],$$

$$\varphi_D(u) = -\mathbf{1}_{\{s<\tau\}}\frac{B(s,T)}{D(s,T)}, \quad \varphi_B(u) = \mathbf{1}_{\{s<\tau\}} \qquad\text{for each } u \in (s,t],$$

$$\varphi_D(u) = 0, \quad \varphi_B(u) = \mathbf{1}_{\{s<\tau\}}\left(1 - \frac{B(s,T)}{D(s,T)}\frac{D(t,T)}{B(t,T)}\right) \quad\text{for each } u \in (t,T].$$

(In particular, if $t = T$, then the last line is void; we rebalance only once at time s.) This is a self-financing simple strategy with initial value $V_\varphi(0) = 0$. Let us consider the terminal value. On $\{\tau \leq s\}$ we have $V_\varphi(T) = 0$. On $\{s < \tau \leq t\}$ we have $D(t,T) = 0$, so $\varphi_B(u) = 1$ and $V_\varphi(T) = 1$, hence $V_\varphi(T) > 0$ with positive probability. And on $\{t < \tau\}$ we have

$$V_\varphi(t)\mathbf{1}_{\{t<\tau\}} = B(t,T) - \frac{B(s,T)}{D(s,T)}D(t,T)$$

$$= e^{-r(T-t)}(1 - e^{g(s)-g(t)}) \geq 0.$$

We can see that the terminal value $V_\varphi(T)$ is always non-negative, and it is strictly positive with positive probability, a contradiction with the No Simple Arbitrage Principle. $\square$

Remark 2.40

The yield on the defaultable bond between times t and T is $\frac{1}{T-t}\ln\frac{1}{D(t,T)}$, and the **credit spread** is

$$s(t,T) = \frac{1}{T-t}\ln\frac{1}{D(t,T)} - r = \frac{1}{T-t}g(t).$$

The fact that g is strictly decreasing implies that $g(t)$ must be positive at any time $t < T$ since $g(T) = 0$, which means that the credit spread is positive.

Let us examine the dynamics of the defaultable bond. While the bond is still alive, that is, while $t < \tau$, we have $D(t,T) = e^{-r(T-t)}e^{-g(t)}$, that is, the bond exhibits exponential growth modified by the factor $e^{-g(t)}$, where $g : [0,T] \to \mathbb{R}$ is a strictly decreasing right-continuous deterministic function such that $g(T) = 0$. This growth terminates at time τ, when the price of the defaultable bond drops to zero. This gives

$$D(t,T) = \mathbf{1}_{\{t<\tau\}}e^{-r(T-t)}e^{-g(t)}. \tag{2.7}$$

The growth of a defaultable bond $D(t,T)$ as compared with a default-free bond $B(t,T)$ is illustrated in Figure 2.2 in the case when $\tau < T$.

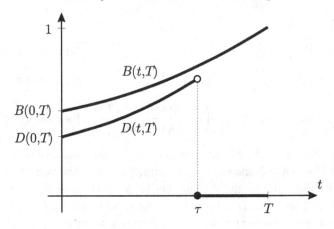

Figure 2.2 Graphs of $D(t, T)$ and $B(t, T)$.

Remark 2.41

It follows from (2.7) that

$$D(t, T) = (1 - I(t))e^{-r(T-t)}e^{-g(t)},$$

where $e^{-r(T-t)}e^{-g(t)}$ is a deterministic function, hence the filtration $(\mathcal{I}_t)_{t \geq 0}$ generated by the default indicator process $I(t) = \mathbf{1}_{\{\tau \leq t\}}$ is the same as that generated by the defaultable bond price process $D(t, T)$.

The behaviour of the defaultable bond given by (2.7) is clearly a necessary condition for the No Simple Arbitrage Principle to hold. It turns out to be also sufficient.

Theorem 2.42

If

$$D(t, T) = \begin{cases} e^{-r(T-t)-g(t)} & \text{for } t < \tau, \\ 0 & \text{for } \tau \leq t, \end{cases}$$

where $g : [0, T] \to \mathbb{R}$ is a positive and strictly decreasing right-continuous function with $g(T) = 0$, then there is no simple arbitrage in the model.

Proof Let $\varphi = (\varphi_B, \varphi_D)$ be a self-financing simple strategy with rebalancing times $0 = S_0 < S_1 < \cdots < S_n = T$ such that $V_\varphi(0) = 0$ and $V_\varphi(T) \neq 0$ on a set of positive probability. We shall show that $V_\varphi(T) > 0$ with positive probability and $V_\varphi(T) < 0$ with positive probability, so it cannot be a simple arbitrage strategy.

Take the smallest integer k such that $V_\varphi(S_k) \neq 0$ on a set of positive probability. Clearly, $k > 1$. The self-financing condition at time S_{k-1} gives

$$0 = V_\varphi(S_{k-1}) = \varphi_B(S_k)B(S_{k-1}, T) + \varphi_D(S_k)D(S_{k-1}, T).$$

On $\{\tau \le S_{k-1}\}$ we therefore have $\varphi_B(S_k) = 0$ because $D(S_{k-1}, T) = 0$. Since $\varphi_B(S_k)$ is $\mathcal{I}_{S_{k-1}}$-measurable, it must be constant on $\{S_{k-1} < \tau\}$, which is an atom in $\mathcal{I}_{S_{k-1}}$. This constant must be non-zero or else $\varphi_B(S_k) = 0$, hence $\varphi_D(S_k) = 0$ and so $V_\varphi(S_k) = 0$ everywhere, contradicting the choice of k. There are two possibilities.

Case 1: $\varphi_B(S_k) > 0$ on $\{S_{k-1} < \tau\}$.

On $\{S_{k-1} < \tau \le S_k\}$ we have $D(S_k, T) = 0$, and it follows that $V_\varphi(S_k) = \varphi_B(S_k)B(S_k, T) > 0$. For all later times the defaultable bonds remain worthless, so the value of the strategy must remain positive on $\{S_{k-1} < \tau \le S_k\}$. In particular, $V_\varphi(T) > 0$ on $\{S_{k-1} < \tau \le S_k\}$, which has positive probability.

On $\{S_k < \tau\}$ the long position $\varphi_B(S_k) > 0$ in non-defaultable bonds is perfectly balanced by the short position $\varphi_D(S_k) = -\frac{B(S_{k-1}, T)}{D(S_{k-1}, T)}\varphi_B(S_k) < 0$ in defaultable bonds so that $V_\varphi(S_{k-1}) = 0$ at time S_{k-1}. Prior to a default the defaultable bond grows faster than the non-defaultable one, so at time S_k the long position will become dominated by the short one, hence $V_\varphi(S_k) < 0$ on $\{S_k < \tau\}$. If $k = n$, then $S_k = T$, so $V_\varphi(T) < 0$ on the set $\{T < \tau\}$ of positive probability. If $k < n$, then by the self-financing condition $\varphi_B(S_{k+1})B(S_k, T) = V_\varphi(S_k) < 0$ on $\{S_k < \tau \le S_{k+1}\}$, so $\varphi_B(S_{k+1}) < 0$ and $V_\varphi(S_{k+1}) = \varphi_B(S_{k+1})B(S_{k+1}, T) < 0$ on $\{S_k < \tau \le S_{k+1}\}$. For all times later than S_{k+1} the defaultable bonds remain worthless, hence the value of the strategy must remain negative on $\{S_k < \tau \le S_{k+1}\}$. In particular, $V_\varphi(T) < 0$ on $\{S_k < \tau \le S_{k+1}\}$, which has positive probability. This completes the argument in Case 1.

Case 2: $\varphi_B(S_k) < 0$ on $\{S_{k-1} < \tau\}$.

Taking the self-financing simple strategy $-\varphi(t) = (-\varphi_B(t), -\varphi_D(t))$ for $t \in [0, T]$ reduces this to Case 1. $\qquad\square$

Remark 2.43

A popular alternative definition of a simple strategy is one where deterministic rebalancing times are replaced by stopping times with respect to the underlying filtration. Here the deterministic rebalancing times $0 = S_0 < S_1 < \cdots < S_n = T$ would be replaced by stopping times $0 = \sigma_0 < \sigma_1 < \cdots < \sigma_n = T$ with respect to the filtration $(\mathcal{I}_t)_{t\ge 0}$. Moreover, for each $k = 1, \ldots, n$ the $\mathcal{I}_{S_{k-1}}$-measurable portfolios $\varphi_B(S_k), \varphi_D(S_k)$ constant over $(S_{k-1}, S_k]$ would be replaced by portfolios $\varphi_B(\sigma_k), \varphi_D(\sigma_k)$ constant over $(\sigma_{k-1}, \sigma_k]$ and measurable with respect to the σ-field $\mathcal{I}_{\sigma_{k-1}}$ that consists of all events $A \subset \Omega$ such that $A \cap \{\sigma_{k-1} \le t\} \in \mathcal{I}_t$ for each $t \ge 0$.

This gives a larger set of simple strategies, hence all properties proved above under the no-simple-arbitrage assumption will also be valid if the alternative definition is adopted. Moreover, the sufficient condition for the No Simple Arbitrage Principle to hold established in Theorem 2.42 can also be shown to be sufficient under the alternative definition of a simple strategy; see Exercise 2.13. The definition of a simple strategy used in this chapter has the advantage of being technically simpler and is sufficient for our purposes.

Exercise 2.11 Let σ be a stopping time with respect to the filtration $(\mathcal{I}_t)_{t\geq 0}$ such that $0 \leq \sigma \leq T$. Show that there is a deterministic constant $c \in [0, T]$ such that $\sigma \geq \tau \wedge c$ and $\sigma = c$ on $\{c < \tau\}$.

Exercise 2.12 Let σ be an $(\mathcal{I}_t)_{t\geq 0}$-stopping time such that $0 \leq \sigma \leq T$. Show that there is a deterministic constant $c \in [0, T]$ such that $\{\sigma < \tau\} = \{c < \tau\}$ and that $\{c < \tau\}$ is an atom in the σ-field $\mathcal{I}_\sigma$ consisting of all events $A \subset \Omega$ such that $A \cap \{\sigma \leq t\} \in \mathcal{I}_t$ for each $t \geq 0$.

Exercise 2.13 Show that the sufficient condition for the No Simple Arbitrage Principle to hold stated in Theorem 2.42 is also sufficient under the alternative definition of a simple strategy indicated in Remark 2.43 with $(\mathcal{I}_t)_{t\geq 0}$-stopping times as rebalancing times.

Risk-neutral probability

We want to explore the possibility of constructing a probability measure Q equivalent to P such that the discounted defaultable bond price $e^{-rt}D(t, T)$ would become a martingale under Q. A necessary condition for a martingale is that the expectation should be constant, which gives

$$\mathbb{E}_Q(e^{-rt}D(t, T)) = D(0, T).$$

Inserting $D(t, T) = \mathbf{1}_{\{t<\tau\}}e^{-r(T-t)-g(t)}$, we get

$$\mathbb{E}_Q(e^{-rt}D(t, T)) = e^{-rT-g(t)}\mathbb{E}_Q(\mathbf{1}_{\{t<\tau\}})$$
$$= e^{-rT-g(t)}Q(t < \tau).$$

For this to be equal to $D(0,T) = e^{-rT-g(0)}$ we would need

$$Q(t < \tau) = e^{-(g(0)-g(t))}$$

for any $t \in [0,T]$.

This allows us to define a probability measure Q on the σ-field $\mathcal{I}_T$. The measure of $\{T < \tau\}$, which is an atom in $\mathcal{I}_T$, should be

$$Q(T < \tau) = e^{-(g(0)-g(T))} = e^{-g(0)}$$

since $g(T) = 0$. On the complement of the atom we have

$$Q(\tau \leq t) = 1 - Q(t < \tau) = 1 - e^{-(g(0)-g(t))}$$

for any $t \in [0,T]$. By Assumption 2.29, $D(t,T) = \mathbf{1}_{\{t<\tau\}}e^{-r(T-t)-g(t)}$ has right-continuous paths. Because $\{t < \tau\}$ has positive probability for each t, it follows that the deterministic function $g(t)$ is right-continuous, so Q can be extended to a measure on the σ-field $\mathcal{I}_T$ generated by sets of the form $\{\tau \leq t\}$ for $t \in [0,T]$.

The next result shows that the construction of Q, even though it involves only the necessary condition for a martingale, ensures that $e^{-rt}D(t,T)$ is indeed a martingale under Q.

Theorem 2.44

The No Simple Arbitrage Principle implies that the discounted defaultable bond prices $e^{-rt}D(t,T)$ for $t \in [0,T]$ follow an $(\mathcal{I}_t)_{t\geq0}$-martingale under the probability measure Q defined on $\mathcal{I}_T$.

Proof We will show that

$$e^{-rt}D(t,T) = e^{-rT}\mathbb{E}_Q(D(T,T)|\mathcal{I}_t) \tag{2.8}$$

for any $t \in [0,T]$. To this end we compute the conditional expectation on the right-hand side using Proposition 2.27, which was proved for P but is in fact valid for any probability measure Q as long as $Q(t < \tau) > 0$. This gives

$$\mathbb{E}_Q(D(T,T)|\mathcal{I}_t) = \mathbf{1}_{\{\tau\leq t\}}\mathbb{E}_Q(\mathbf{1}_{\{T<\tau\}}|\sigma(\tau)) + \mathbf{1}_{\{t<\tau\}}\frac{\mathbb{E}_Q(\mathbf{1}_{\{t<\tau\}}\mathbf{1}_{\{T<\tau\}})}{Q(t<\tau)}$$

$$= \mathbf{1}_{\{\tau\leq t\}}\mathbf{1}_{\{T<\tau\}} + \mathbf{1}_{\{t<\tau\}}\frac{Q(T<\tau)}{Q(t<\tau)}$$

$$= 0 + \mathbf{1}_{\{t<\tau\}}\frac{e^{-g(0)+g(T)}}{e^{-g(0)+g(t)}}$$

$$= \mathbf{1}_{\{t<\tau\}}e^{-g(t)},$$

which after multiplying by e^{-rT} gives $\mathbf{1}_{\{t<\tau\}}e^{-rT-g(t)}$. Finally, we use the fact that $D(t, T) = \mathbf{1}_{\{t < \tau\}} e^{-r(T-t)-g(t)}$ to conclude that the right-hand side of (2.8) is equal to $e^{-rt} D(t, T)$.   □

Theorem 2.45

Under the No Simple Arbitrage Principle, the probability Q on $\mathcal{I}_T$ under which $e^{-rT} D(t, T)$ is an $(\mathcal{I}_t)_{t \geq 0}$-martingale is unique.

Proof   Suppose that $\bar{Q}$ is another probability measure on $\mathcal{I}_T$ under which $e^{-rT} D(t, T)$ is an $(\mathcal{I}_t)_{t \geq 0}$-martingale. It follows that

$$D(0, T) = e^{-rt} \mathbb{E}_{\bar{Q}}(D(t, T))$$
$$= e^{-rt} \mathbb{E}_{\bar{Q}}(\mathbf{1}_{\{t < \tau\}} e^{-r(T-t)-g(t)}) = e^{-rT-g(t)} \bar{Q}(t < \tau).$$

Similarly, since $e^{-rT} D(t, T)$ is an $(\mathcal{I}_t)_{t \geq 0}$-martingale under Q,

$$D(0, T) = e^{-rt} \mathbb{E}_Q(D(t, T))$$
$$= e^{-rt} \mathbb{E}_Q(\mathbf{1}_{\{t < \tau\}} e^{-r(T-t)-g(t)}) = e^{-rT-g(t)} Q(t < \tau).$$

Hence

$$\bar{Q}(t < \tau) = Q(t < \tau)$$

for each $t \in [0, T]$, so $\bar{Q} = Q$.   □

For Q to be a risk-neutral probability (a martingale measure) it has to be equivalent to the real-life probability P. For this to hold, the distributions of τ under P and Q need to be equivalent. This, however, may not be true in general, as shown in the next example. Here we use the fact that there is a function $h : [0, 1] \rightarrow [0, 1]$ strictly increasing, continuous, and such that $h'(x) = 0$ for each x in a set of Lebesgue measure 1, with $h(0) = 0$ and $h(1) = 1$.[1]

Example 2.46

On $\Omega = (0, \infty)$ with the σ-field of Borel sets and probability P such that $P(a, b] = e^{-\lambda a} - e^{-\lambda b}$ for any $0 < a \leq b$, where λ is a positive constant, take $\tau(x) = x$ for each $x \in (0, \infty)$. By Theorem 2.42, the model with $T = 1$ and

$$D(t, T) = \mathbf{1}_{\{t < \tau\}} e^{-r(T-t)-g(t)}$$

with $g(t) = h(T) - h(t)$ for any $t \in [0, T]$ satisfies the No Simple Arbitrage Principle. The corresponding probability measure Q given by Theorem 2.44 is not equivalent to P since the latter has a density with respect

---

[1]   A construction of such a function can be found, for example, on pp. 427–429 of P. Billingsley, *Probability and Measure*, 2nd edition, John Wiley & Sons, 1986.

to the Lebesgue measure. By the uniqueness of Q, there is no equivalent risk-neutral probability in this example.

In order to obtain a risk-neutral probability we have to go one step further and impose the following condition, which excludes some possible pathologies.

Assumption 2.47
The real-life probability measure P is equivalent to the unique probability measure Q on I_T such that $e^{-rt}D(t,T)$ is an $(I_t)_{t\geq 0}$-martingale under Q.

Note that τ is not measurable with respect to the σ-field I_T, on which Q is defined. This is because sets of the form $\{\tau \leq t\}$ do not belong to I_T for any $t > T$. Because of this we cannot consider the distribution function of τ under Q. On the other hand,

$$\bar{\tau} = \begin{cases} \tau & \text{on } \{\tau \leq T\} \\ \infty & \text{on } \{T < \tau\} \end{cases}$$

is I_T-measurable. Its probability distribution under Q is well defined and can be characterised by any of the following functions:
- the distribution function $F_Q(t) = Q(\bar{\tau} \leq t)$;
- the survival probability function $G_Q(t) = Q(t < \bar{\tau})$;
- the hazard function $\Gamma_Q(t) = -\ln G_Q(t)$.

Exercise 2.14 Let $\lambda > 0$ and let $g(t) = \lambda(T - t)$ for each $t \in [0, T]$. Compute F_Q, G_Q, and Γ_Q.

Proposition 2.48
For any $t \in [0, T]$

$$\Gamma_Q(t) = g(0) - g(t).$$

Proof By the definition of the hazard function,

$$e^{-\Gamma_Q(t)} = Q(t < \bar{\tau}) = Q(t < \tau) = e^{-g(0)+g(t)}$$

as established above. □

In particular, $\Gamma_Q(T) = g(0)$, so

$$g(t) = \Gamma_Q(T) - \Gamma_Q(t)$$

and the bond prices take the form

$$D(t, T) = \mathbf{1}_{\{t < \tau\}} e^{-r(T-t) + \Gamma_Q(t) - \Gamma_Q(T)}$$

for each $t \in [0, T]$.

Finally, we look at the question of existence of a density for the distribution of $\tilde{\tau}$ and the hazard rate under the probability measure Q. It follows from Assumption 2.47 that the distributions of $\tilde{\tau}$ under P and Q are equivalent. By the Radon–Nikodym theorem (see [PF]), there exists a Borel function χ such that

$$F_Q(t) = Q(\tilde{\tau} \le t) = \int_0^{t \wedge T} \chi(u) f_P(u) du$$

for each $t \in \mathbb{R}$, which implies that $f_Q(t) = \chi(t) f_P(t)$ is the density of $\tilde{\tau}$ under Q. As a consequence, by a similar argument as in Proposition 2.13, the hazard function can be written as

$$\Gamma_Q(t) = \int_0^{t \wedge T} \gamma_Q(u) du,$$

where

$$\gamma_Q(t) = \frac{f_Q(t)}{1 - F_Q(t)}$$

is the hazard rate under Q.

3

Defaultable bond pricing with hazard function

In Chapter 2 we investigated the existence of a risk-neutral probability measure in a market model admitting no simple arbitrage strategies. Here we make a fresh start by adopting the existence of a risk-neutral probability measure as the main assumption, and derive a theory based on this. In particular, we give an alternative derivation of the formula for defaultable bond prices; see Theorem 3.11.

Recall that the time of default is a random variable $\tau > 0$ defined on a probability space $(\Omega, \mathcal{F}, P)$, with strictly increasing distribution function F_P, which has a density f_P. The default indicator process $I(t) = \mathbf{1}_{\{t \leq \tau\}}$ generates the filtration $(\mathcal{I}_t)_{t \geq 0}$. The market consists of two assets, a default-free bond $B(t, T) = e^{-r(T-t)}$ and a defaultable bond $D(t, T)$ with terminal payoff $D(T, T) = \mathbf{1}_{\{T < \tau\}}$.

Our aim is to develop pricing methods for various derivative securities based on replicating their payoffs by means of these two bonds, and to establish pricing formulae involving the expectation under the risk-neutral probability.

The No Simple Arbitrage Principle allowed us to partially solve the problem of finding the risk-neutral probability. In Chapter 2 we constructed a probability measure Q on the σ-field $\mathcal{I}_T$, where T is the maturity date of the underlying bonds. The equivalence of Q and the real-life probability P could not be established (see Example 2.46) and had to be assumed. The dependence of Q on T was not explored. This motivates the following assumption.

Assumption 3.1
There is a probability measure Q defined on the σ-field $\sigma(\tau)$ and equivalent to P such that the discounted defaultable bond prices $e^{-rt}D(t,T)$ form an $(\mathcal{I}_t)_{t\geq 0}$-martingale under Q. We refer to Q as the **risk-neutral probability** or **martingale measure**.

We begin with a simple consequence of the equivalence of P and Q.

Proposition 3.2
The random variable τ has a density f_Q under Q. The distribution function F_Q of τ is strictly increasing on $[0,\infty)$, and $F_Q(t) < 1$ for each $t \in \mathbb{R}$.

Proof By Assumption 2.7, τ has a density f_P under P. Because Q and P are equivalent probability measures, by the Radon–Nikodym theorem (see [PF]) there exists a Borel function χ such that, for each $t \geq 0$,

$$F_Q(t) = Q(\tau \leq t) = \int_0^t \chi(u)f_P(u)du,$$

hence $f_Q(t) = \chi(t)f_P(t)$ is a density of τ under Q.

Next, take any $0 \leq s < t$. By Assumption 2.5, we know that $P(s < \tau \leq t) > 0$. Since Q and P are equivalent probability measures, it follows that $Q(s < \tau \leq t) > 0$. This means that F_Q is a strictly increasing function on $[0,\infty)$. It follows immediately that $F_Q(t) < 1$ for each t. $\square$

In what follows we are going to drop the subscript Q at the expectation and any other quantity like F, G, Γ, f, γ which depends on the choice of the probability measure. We shall be working primarily with the risk-neutral probability Q. From now on

$$\mathbb{E}(X) = \int_\Omega X dQ,$$

$$F(t) = Q(\tau \leq t) = \int_0^t f(u)du,$$

$$G(t) = Q(t < \tau) = \int_t^\infty f(u)du,$$

$$\Gamma(t) = -\ln Q(t < \tau) = \int_0^t \gamma(u)du$$

for any $t \in \mathbb{R}$.

3.1 Initial prices and calibration

We first compute the price of the defaultable bond at time 0. Since the process $e^{-rt}D(t, T)$ is a martingale under Q,

$$D(0, T) = e^{-rT}\mathbb{E}(D(T, T)).$$

Given that $D(T, T) = \mathbf{1}_{\{T < \tau\}}$, we obtain

$$D(0, T) = e^{-rT}Q(T < \tau).$$

Remark 3.3

Note that, in general, the risk-neutral survival probability $Q(T < \tau)$ implied by the bond prices does not agree with the historical data concerning the default time, which is related to the real-life probability. Even if we have sufficient statistical data to estimate $P(T < \tau)$, we cannot expect that $e^{-rT}P(T < \tau)$ will agree with the bond prices. This is a common feature of all classical theories of mathematical finance.

If the time 0 prices of bonds with all maturities are known, the distribution of τ under Q can be obtained as

$$F(T) = 1 - e^{rT}D(0, T)$$

for any $T > 0$. The hazard function satisfies $Q(\tau > T) = e^{-\Gamma(T)}$, so

$$D(0, T) = e^{-rT - \Gamma(T)} = e^{-\int_0^T (r + \gamma(t))dt}.$$

The following proposition is an immediate consequence of this.

Proposition 3.4

The time 0 bond prices $D(0, T)$ imply the following values for the hazard function and, provided the partial derivative exists, also for the hazard rate:

$$\Gamma(T) = -\ln D(0, T) - rT,$$

$$\gamma(T) = -\frac{1}{D(0, T)}\frac{\partial}{\partial T}D(0, T) - r.$$

Remark 3.5

Recall that the **credit spread** is defined as

$$s(T) = k_D - r, \quad \text{where } k_D = -\frac{1}{T}\ln D(0, T),$$

so $s(T) = \frac{\Gamma(T)}{T}$. We saw in Proposition 1.19 that the short-term credit spread $\lim_{T \searrow 0} s(T)$ is zero in the Merton model, which is not a feature of real markets. One of the advantages of the reduced-form approach is that the short-term spread may be positive. It is equal to the right-sided derivative of $\Gamma(T)$ at $T = 0$, that is, to $\gamma(0) = \lim_{T \searrow 0} \frac{\Gamma(T)}{T}$ (if the limit exists). This gives some flexibility to build a model such that the short-term credit spread is positive, consistent with the data.

Example 3.6
Suppose that τ is exponentially distributed under Q with intensity $\lambda > 0$. Then the hazard rate is constant and equal to λ, and $\Gamma(T) = \lambda T$, hence the short-term credit spread is λ.

In practice, the bond prices are known for a finite set $0 < T_1 < \cdots < T_N$ of maturity dates only. The bond prices $D(0, T)$ can be interpolated, for example, by linear functions between any two dates T_{k-1}, T_k for $k = 1, \ldots, N$, with $D(0, T_0) = 1$ at $T_0 = 0$.

Exercise 3.1   Suppose that $D(0, \frac{1}{2}) = 0.9268$, $D(0, \frac{3}{4}) = 0.8487$, $D(0, 1) = 0.7635$, and $r = 5\%$. Extend the bond prices to a piecewise linear function $D(0, T)$ for $T \in [0, 1]$ and compute the corresponding hazard rate.

If we know the bond prices for some maturities only, we can estimate the parameters of the probability distribution of τ if some specific probability distribution is assumed.

Exercise 3.2   Suppose that $r = 5\%$ and $D(0, \frac{1}{2}) = 0.9133$, and assume that τ has the exponential distribution with parameter λ under the risk-neutral probability Q. Find the value of λ and compute $D(0, \frac{1}{4})$.

Example 3.7
Suppose that Γ is quadratic, $\Gamma(T) = aT^2 + bT$ (recall that $\Gamma(0) = 0$)), and two bonds with maturities T_1, T_2 are traded. Their initial prices $D(0, T_1)$ and $D(0, T_2)$ determine Γ via a system of two equations in two variables a

and b, namely

$$aT_1^2 + bT_1 = -\ln D(0, T_1) - rT_1,$$
$$aT_2^2 + bT_2 = -\ln D(0, T_2) - rT_2.$$

Exercise 3.3   Given $D(0, \frac{1}{2}) = 0.8679$, $D(0, 1) = 0.7055$, and $r = 5\%$, compute $D(0, \frac{3}{4})$ assuming that Γ is quadratic.

Remark 3.8

The above examples and exercises are somewhat naive. In practice, bond prices are usually known for more than just one or two maturity dates, hence the above problems would become overdetermined, admitting no exact solution. In such circumstances we are typically looking for parameter values that give the best match with the observed bond prices in a certain sense (e.g. least square optimisation) rather than an exact match.

Example 3.9

Suppose that the bond prices $D(0, T_k)$ are known for finitely many maturity dates $0 < T_1 < \cdots < T_N$. Assume Γ to be linear between any two dates T_{k-1}, T_k for $k = 1, \ldots, N$ (with $T_0 = 0$). This means that the hazard rate γ is constant on $[0, T_1], (T_1, T_2], \ldots, (T_{N-1}, T_N]$, with values $\gamma_1, \gamma_2, \ldots, \gamma_N$ such that

$$-\ln D(0, T_1) - rT_1 = \Gamma(T_1) = \gamma_1 T_1,$$
$$-\ln D(0, T_2) - rT_2 = \Gamma(T_2) = \gamma_1 T_1 + \gamma_2 (T_2 - T_1),$$
$$\vdots$$
$$-\ln D(0, T_N) - rT_N = \Gamma(T_N) = \gamma_1 T_1 + \gamma_2 (T_2 - T_1) + \cdots + \gamma_N (T_N - T_{N-1}).$$

It follows that

$$\gamma_1 = \frac{1}{T_1} \ln \frac{1}{D(0, T_1)} - r,$$

$$\gamma_2 = \frac{1}{T_2 - T_1} \ln \frac{D(0, T_1)}{D(0, T_2)} - r,$$

$$\vdots$$

$$\gamma_N = \frac{1}{T_N - T_{N-1}} \ln \frac{D(0, T_{N-1})}{D(0, T_N)} - r.$$

Exercise 3.4 Given $D(0, \frac{1}{2}) = 0.85$, $D(0, 1) = 0.8$, and $r = 5\%$, compute $D(0, \frac{3}{4})$ assuming that the hazard rate γ is constant on $[0, \frac{1}{2}]$ and $(\frac{1}{2}, 1]$.

Exercise 3.5 Let $D(0, \frac{1}{2}) = 0.9037$, $D(0, \frac{3}{4}) = 0.8609$, $D(0, 1) = 0.7724$, and $r = 5\%$. Assuming that the hazard rate γ is constant on $[0, \frac{1}{2}], (\frac{1}{2}, \frac{3}{4}], (\frac{3}{4}, 1]$, compute the values it takes on these intervals.

3.2 Process of prices

Now we turn to the task of finding the form of the process $D(t, T)$. According to Assumption 3.1, the discounted prices $e^{-rt}D(t, T)$ form a martingale under Q, so

$$D(t, T) = e^{-r(T-t)}\mathbb{E}(\mathbf{1}_{\{T < \tau\}}|\mathcal{I}_t).$$

This conditional expectation can be computed with the aid of the results proved in Chapter 2. We begin with an example.

Example 3.10
Let τ be exponentially distributed with parameter $\lambda > 0$ under the risk-neutral probability Q. For better clarity we may assume that $\Omega = [0, 1]$

with Q the Lebesgue measure and $\tau(\omega) = -\frac{1}{\lambda}\ln(1-\omega)$. Take any $t \in [0,T]$. Because $\mathbb{E}(\mathbf{1}_{\{T<\tau\}}|\mathcal{I}_t)$ is $\mathcal{I}_t$-measurable, we know that it is constant on the event $\{t < \tau\}$, say $\mathbb{E}(\mathbf{1}_{\{T<\tau\}}|\mathcal{I}_t) = c_t$, and

$$\int_{\{t<\tau\}} \mathbf{1}_{\{T<\tau\}} dQ = \int_{\{t<\tau\}} c_t \, dQ.$$

The right-hand side is $c_t Q(t < \tau) = c_t e^{-\lambda t}$. On the left-hand side we have $Q(T < \tau) = e^{-\lambda T}$ since $\{T < \tau\} \subset \{t < \tau\}$. Hence

$$c_t = e^{-\lambda(T-t)}.$$

On the event $\{\tau \leq t\}$ the random variable $\mathbf{1}_{\{T<\tau\}}$ is zero, so

$$D(t,T) = \mathbf{1}_{\{t<\tau\}} e^{-(r+\lambda)(T-t)}.$$

The process $D(t,T)$ starts from the initial value $D(0,T) = e^{-(r+\lambda)T}$ and grows exponentially at the rate $r + \lambda$ up to time τ, when it drops down to zero.

We generalise this example to find a formula for the bond price process for any τ.

Theorem 3.11

The process of bond prices has the form

$$D(t,T) = \mathbf{1}_{\{t<\tau\}} e^{-r(T-t)} e^{-(\Gamma(T)-\Gamma(t))} \tag{3.1}$$

for each $t \in [0,T]$.

Proof We apply Corollary 2.26 to get

$$D(t,T) = e^{-r(T-t)} \mathbb{E}\left(\mathbf{1}_{\{T<\tau\}}|\mathcal{I}_t\right) = e^{-r(T-t)} \mathbf{1}_{\{t<\tau\}} e^{-(\Gamma(T)-\Gamma(t))}.$$

$\square$

The process $D(t,T)$ is quite simple. On the event $\{\tau \leq t\}$ the value of the defaultable bond at time t is zero. The bond has ceased to exist at time τ, and there is no recovery payment. On $\{t < \tau\}$ the bond price is given by a deterministic function. This motivates the next definition, which will later be extended to more general situations; cf. Definition 6.2.

Definition 3.12

If X is an $(\mathcal{I}_t)_{t\geq 0}$-adapted process, then a deterministic function $\hat{X}$ such that, for all t,

$$X(t)\mathbf{1}_{\{t<\tau\}} = \hat{X}(t)\mathbf{1}_{\{t<\tau\}} \qquad (3.2)$$

is called the **pre-default value** corresponding to X.

When $X(t)$ is integrable, the pre-default value exists and is uniquely determined by the defining equality (3.2). We can take the expectation with respect to Q on both sides of the equality and divide by $Q(t < \tau)$ to get

$$\hat{X}(t) = \frac{\mathbb{E}(X(t)\mathbf{1}_{\{t<\tau\}})}{Q(t < \tau)} = \mathbb{E}(X(t)|t < \tau).$$

Furthermore, if $e^{-rt}X(t)$ is a martingale, then $\mathbb{E}(e^{-rT}X(T)|\mathcal{I}_t) = e^{-rt}X(t)$ is constant on $\{t < \tau\}$ and equal to $e^{-rt}\hat{X}(t)$, where $\hat{X}(t) = e^{-r(T-t)}\mathbb{E}(X(T)|t < \tau)$.

At time 0 we have $\hat{X}(0) = X(0)$ almost surely since $Q(\tau > 0) = 1$. (The property $P(\tau > 0) = 1$ is assumed for the real-life probability P, but preserved for Q, which is equivalent to P.)

There are several equivalent forms for the pre-default value of a defaultable bond:

$$\hat{D}(t,T) = e^{-r(T-t)}e^{-(\Gamma(T)-\Gamma(t))} = e^{-r(T-t)}\frac{G(T)}{G(t)} = e^{-r(T-t)}\frac{1 - F(T)}{1 - F(t)}.$$

Example 3.13
If τ is exponentially distributed under Q, then the pre-default value of the defaultable bond is

$$\hat{D}(t,T) = e^{-(r+\lambda)(T-t)},$$

which can be viewed as the time t value of a unit amount due at time T discounted by using $r + \lambda$ as if it were the interest rate.

The bond price can also be expressed in terms of the hazard rate,

$$D(t,T) = \mathbf{1}_{\{t<\tau\}}e^{-r(T-t)}e^{-\int_t^T \gamma(s)ds} = \mathbf{1}_{\{t<\tau\}}e^{-\int_t^T (r+\gamma(s))ds},$$

which gives a simple and intuitive formula for the pre-default value, namely

$$\hat{D}(t,T) = e^{-\int_t^T (r+\gamma(s))ds}.$$

As a result, $\hat{D}(t,T)$ satisfies

$$\hat{D}(t,T) - \hat{D}(0,T) = \int_0^t r\hat{D}(s,T)ds + \int_0^t \gamma(s)\hat{D}(s,T)ds. \qquad (3.3)$$

This equation will be used in the proof of Theorem 3.21.

Exercise 3.6 Find the function $\hat{D}(t, 1)$ if $D(0, \frac{1}{2}) = 0.85$, $D(0, 1) = 0.8$, $r = 5\%$, and γ is constant on $[0, \frac{1}{2}]$ and $(\frac{1}{2}, 1]$.

3.3 Recovery schemes

The defaultable bond considered above pays nothing when a default occurs before or on the maturity date, that is, when $\tau \leq T$. However, in reality, in such cases bond holders often receive partial compensation, a so-called **recovery payment**, which can fall into the following categories.

(i) *Recovery at maturity.* This is when defaultable bond holders receive a recovery payment at time T if a default occurs before or at time T.

  (a) *Constant recovery.* The simplest case is when the recovery payment is a constant amount δ known in advance and paid on the maturity date T if $\tau \leq T$. Hence the bond pays $\mathbf{1}_{\{\tau > T\}} + \delta \mathbf{1}_{\{\tau \leq T\}}$ at maturity.

  (b) *General recovery.* A constant recovery payment is unrealistic since the amount of money generated by liquidating the company's assets following bankruptcy can be difficult to predict in advance. Given the information available at time T, represented by the σ-field $\mathcal{I}_T$, the most general recovery payment at maturity can be described by an $\mathcal{I}_T$-measurable random variable. According to Proposition 2.22 and Remark 2.23, such a random variable can be expressed as $\phi(\tau)$ for some Borel function ϕ. At maturity the bond pays $\mathbf{1}_{\{\tau > T\}} + \phi(\tau)\mathbf{1}_{\{\tau \leq T\}}$.

(ii) *Recovery at default.* The bond holders would not normally have to wait until the maturity time T to receive the money obtained from liquidation when a company goes bankrupt before the bonds mature. Market practice suggests a number of different schemes for a recovery payment due at the default time τ in the case when $\tau \leq T$. In our simple market model such recovery payment can only be invested in the default-free bond, as this is the only asset surviving beyond time τ.

  (a) *Constant recovery.* A fixed amount δ is paid at the default time τ if $\tau \leq T$. Invested in the default-free bond, it will become $\delta e^{r(T-\tau)}$ on the maturity date T. This case falls under (i)(b) above. A restriction on δ is needed so that the terminal value does not exceed the face value of the bond. Namely, $\delta e^{r(T-\tau)} \leq 1$, so the constant δ should satisfy $\delta \leq e^{-rT}$.

(b) *Fractional recovery of treasury value.* By the treasury value we mean the price of the default-free bond. Suppose that $\delta \in [0, 1]$ represents a fraction of this price paid at default, so the bond holder receives $\delta B(\tau, T) = \delta e^{-r(T-\tau)}$ at time τ. Invested in the default-free bond, this gives constant δ at time T, and therefore falls under (i)(a) and of course also under (i)(b).

(c) *Fractional recovery of market value.* By the market value we mean the value of the defaultable bond immediately prior to τ (at time τ we have $D(\tau, T) = 0$), that is, the pre-default value. Let $\delta \in [0, 1]$ represent a fraction of this value such that the recovery payment will be

$$\lim_{t \nearrow \tau} \delta D(t, T) = \delta \hat{D}(\tau, T).$$

Form (3.2) we have $\delta \hat{D}(\tau, T) = \delta e^{-r(T-\tau)} e^{-(\Gamma(T)-\Gamma(\tau))}$. Invested in the default-free bond, it becomes $\delta e^{-(\Gamma(T)-\Gamma(\tau))}$ at maturity. This case is also covered by (i)(b). No additional restriction on δ is needed since $\Gamma(T) \geq \Gamma(\tau)$.

In general, a recovery payment at the time of default takes the form $\eta(\tau)$ for some Borel function η. For example, in the three cases above,

$$\eta(t) = \begin{cases} \delta & \text{for (ii)(a),} \\ \delta e^{-r(T-t)} & \text{for (ii)(b),} \\ \delta e^{-r(T-t)} e^{-(\Gamma(T)-\Gamma(t))} & \text{for (ii)(c).} \end{cases}$$

When invested in the default-free bond at time τ, the recovery payment $\eta(\tau)$ will grow to become $e^{r(T-\tau)}\eta(\tau)$ on the maturity date T, hence this is also covered under case (i)(b) with $\phi(t) = e^{r(T-t)}\eta(t)$.

The payoff at maturity for a defaultable bond with a recovery payment will be denoted by $h(\tau)$, related to the recovery payment $\phi(\tau)$ at maturity or $\eta(\tau)$ at default by

$$\begin{aligned} h(t) &= \mathbf{1}_{\{T<\tau\}} + \mathbf{1}_{\{\tau \leq T\}}\phi(t) \\ &= \mathbf{1}_{\{T<\tau\}} + \mathbf{1}_{\{\tau \leq T\}}e^{r(T-t)}\eta(t). \end{aligned}$$

It will be convenient to treat all such cases uniformly by considering a general security with payoff at maturity of the form $h(\tau)$ for a given Borel function h. As we know, such payoffs cover all $\mathcal{I}_T$-measurable random variables. So, if we succeed, the problem of pricing general European contingent claims will also be solved.

We therefore set the following task: find the process of prices of a security paying $h(\tau)$ at time T by means of replicating this payoff by the basic assets $B(t, T)$ and $D(t, T)$ in our market model, and represent this process of prices in terms of conditional expectations of the discounted payoff with respect to the filtration $(I_t)_{t \geq 0}$ under the risk-neutral probability Q.

3.4 Martingale properties

Our interest in martingale properties is due to the fact that we want to replicate and price derivative securities with general payoffs of the form $h(\tau)$ and exercise time T. In particular, the discounted values of replicating strategies should be martingales under the risk-neutral probability Q. They will be represented as stochastic integrals with respect to certain driving processes. To achieve the martingale property the driving processes should be martingales themselves.

A natural candidate for such a driving process would be $I(t)$. However, it is not a martingale as its expectation

$$\mathbb{E}(I(t)) = Q(\tau \leq t) = F(t)$$

is increasing, while a necessary condition for a martingale is that the expectation should be constant. Nonetheless, after subtracting a suitable process, called a **compensator** of $I(t)$, we are going to obtain the martingale property. We seek a compensator process that is $(I_t)_{t \geq 0}$-adapted and has non-decreasing continuous trajectories.

Example 3.14
A **Poisson process** with parameter $\lambda > 0$ is an integer-valued stochastic process $N(t), t \geq 0$, with right-continuous trajectories and independent increments such that $N(0) = 0$, and for any $0 \leq s < t$ the increment $N(t) - N(s)$ has the Poisson distribution (see [PF]) with parameter $\lambda(t - s)$.

It is not hard to show that $N(t) - \lambda t$ is a martingale with respect to the filtration $(\mathcal{N}_t)_{t \geq 0}$ generated by the Poisson process (Exercise 3.7), and that the time of the first jump of the Poisson process, defined as

$$\tau = \inf \{t \geq 0 : N(t) > 0\},$$

has the exponential distribution with parameter λ (Exercise 3.8).

Observe that $N(t \wedge \tau) = I(t)$. It is a known fact that a stopped martingale is a martingale (see [SCF]). It follows that

$$N(t \wedge \tau) - \lambda(t \wedge \tau) = I(t) - \lambda(t \wedge \tau)$$

is an $(\mathcal{N}_t)_{t \geq 0}$-martingale. Clearly, $\mathcal{I}_t \subset \mathcal{N}_t$ for all $t \geq 0$. Since $I(s) - \lambda(s \wedge \tau)$ is $\mathcal{I}_s$-measurable, we have

$$\begin{aligned}
\mathbb{E}(I(t) - \lambda(t \wedge \tau)|\mathcal{I}_s) &= \mathbb{E}(\mathbb{E}(I(t) - \lambda(t \wedge \tau)|\mathcal{N}_s)|\mathcal{I}_s) \\
&= \mathbb{E}(I(s) - \lambda(s \wedge \tau)|\mathcal{I}_s) \\
&= I(s) - \lambda(s \wedge \tau),
\end{aligned}$$

so $I(t) - \lambda(t \wedge \tau)$ is also an $(\mathcal{I}_t)_{t \geq 0}$-martingale. Hence $\lambda(t \wedge \tau)$ is a compensator of $I(t)$, adapted to the filtration $(\mathcal{I}_t)_{t \geq 0}$ and with non-decreasing continuous trajectories.

Exercise 3.7 Show that $N(t) - \lambda t$ is a martingale with respect to the filtration $(\mathcal{N}_t)_{t \geq 0}$ generated by the Poisson process.

Exercise 3.8 Show that the time of the first jump of the Poisson process is an exponentially distributed random variable with parameter λ.

The fact that $\lambda(t \wedge \tau)$ is a compensator of $I(t)$ is true not just for the time of the first jump of a Poisson process, but indeed for any exponentially distributed random variable τ.

Exercise 3.9 Suppose that τ is an exponentially distributed random variable with parameter $\lambda > 0$. Show that $I(t) - \lambda(t \wedge \tau)$ is a martingale by verifying directly that for any $0 \leq s < t$

$$\mathbb{E}(I(t) - \lambda(t \wedge \tau)|\mathcal{I}_s) = I(s) - \lambda(s \wedge \tau),$$

and conclude that $\lambda(t \wedge \tau)$ is a compensator of $I(t)$.

For an exponentially distributed τ we have $\Gamma(t) = \lambda t$, hence the compensator of $I(t)$ can be written as $\lambda(t \wedge \tau) = \Gamma(t \wedge \tau)$. This suggests the following general form for an adapted non-decreasing continuous compensator of $I(t)$.

Theorem 3.15
For any random variable τ satisfying the assumptions in Chapter 2, $\Gamma(t \wedge \tau)$ is a compensator of $I(t)$. That is, the **compensated indicator** *process*

$$M(t) = I(t) - \Gamma(t \wedge \tau)$$

is a martingale with respect to the filtration $(\mathcal{I}_t)_{t \geq 0}$.

Proof Let $0 \leq s < t$. By Proposition 2.27 applied to the risk-neutral probability Q, we have

$$\mathbb{E}(I(t)|\mathcal{I}_s) = \mathbb{E}(\mathbf{1}_{\{\tau \leq t\}}|\mathcal{I}_s)$$

$$= \mathbf{1}_{\{\tau \leq s\}}\mathbb{E}(\mathbf{1}_{\{\tau \leq t\}}|\sigma(\tau)) + \mathbf{1}_{\{s < \tau\}}\frac{\mathbb{E}(\mathbf{1}_{\{s < \tau\}}\mathbf{1}_{\{\tau \leq t\}})}{Q(s < \tau)}$$

$$= \mathbf{1}_{\{\tau \leq s\}}\mathbf{1}_{\{\tau \leq t\}} + \mathbf{1}_{\{s < \tau\}}\frac{Q(s < \tau \leq t)}{Q(s < \tau)}$$

$$= I(s) + \mathbf{1}_{\{s < \tau\}}\frac{G(s) - G(t)}{G(s)}.$$

On the other hand, using Proposition 2.27 once again, we have

$$\mathbb{E}(\Gamma(t \wedge \tau)|\mathcal{I}_s) = \mathbf{1}_{\{\tau \leq s\}}\mathbb{E}(\Gamma(t \wedge \tau)|\sigma(\tau)) + \mathbf{1}_{\{s < \tau\}}\frac{\mathbb{E}(\mathbf{1}_{\{s < \tau\}}\Gamma(t \wedge \tau))}{Q(s < \tau)}.$$

The expectations on the right-hand side can be written as

$$\mathbb{E}(\Gamma(t \wedge \tau)|\sigma(\tau)) = \Gamma(t \wedge \tau)$$

since $\Gamma(t \wedge \tau)$ is $\sigma(\tau)$-measurable, and

$$\mathbb{E}(\mathbf{1}_{\{s < \tau\}}\Gamma(t \wedge \tau)) = \int_s^\infty \Gamma(t \wedge u) f(u)du$$

$$= \int_s^t \Gamma(u)f(u)du + \Gamma(t) \int_t^\infty f(u)du$$

$$= \int_s^t \Gamma(u)f(u)du + \Gamma(t)G(t)$$

$$= \Gamma(s)G(s) + G(s) - G(t),$$

where in the last equality we use the integration-by-parts formula

$$\Gamma(t)G(t) - \Gamma(s)G(s) = \int_s^t \gamma(u)G(u)du - \int_s^t \Gamma(u)f(u)du$$

$$= \int_s^t f(u)du - \int_s^t \Gamma(u)f(u)du$$

$$= G(s) - G(t) - \int_s^t \Gamma(u)f(u)du.$$

Altogether, this gives

$$\mathbb{E}\left(\Gamma(t \wedge \tau)|\mathcal{I}_s\right) = \mathbf{1}_{\{\tau \le s\}}\Gamma(t \wedge \tau) + \mathbf{1}_{\{s<\tau\}}\frac{\Gamma(s)G(s) + G(s) - G(t)}{G(s)}$$

$$= \mathbf{1}_{\{\tau \le s\}}\Gamma(\tau) + \mathbf{1}_{\{s<\tau\}}\Gamma(s) + \mathbf{1}_{\{s<\tau\}}\frac{G(s) - G(t)}{G(s)}$$

$$= \Gamma(s \wedge \tau) + \mathbf{1}_{\{s<\tau\}}\frac{G(s) - G(t)}{G(s)}.$$

It follows that

$$\mathbb{E}\left(I(t) - \Gamma(t \wedge \tau)|\mathcal{I}_s\right) = I(s) - \Gamma(s \wedge \tau),$$

so $I(t) - \Gamma(t \wedge \tau)$ is indeed a martingale with respect to the filtration $(\mathcal{I}_t)_{t \ge 0}$. Hence $\Gamma(t \wedge \tau)$ is a compensator of $I(t)$. $\qquad \square$

Exercise 3.10 Compute the expectation $\mathbb{E}(\Gamma(t \wedge \tau))$ to show that it is equal to $F(t)$.

Corollary 3.16
The compensator $\Gamma(t \wedge \tau)$ of the indicator process $I(t)$ can be written as

$$\Gamma(t \wedge \tau) = \int_0^t (1 - I(u))\,\gamma(u)du,$$

where γ is the hazard rate.

Proof By the definition of γ,

$$\Gamma(t \wedge \tau) = \int_0^{t \wedge \tau} \gamma(u)du$$

$$= \int_0^t \mathbf{1}_{\{u < \tau\}} \gamma(u)du$$

$$= \int_0^t \left(1 - \mathbf{1}_{\{\tau \le u\}}\right) \gamma(u)du$$

$$= \int_0^t \left(1 - I(u)\right) \gamma(u)du.$$

$\square$

Exercise 3.11 Show that the compensator $\Gamma(t \wedge \tau)$ of the indicator process $I(t)$ can be written as

$$\Gamma(t \wedge \tau) = \int_0^{t \wedge \tau} \frac{1}{1 - F(u)} f(u)du = \int_0^t \frac{1 - I(u)}{1 - F(u)} f(u)du.$$

We also consider a compensator in multiplicative form for the process $1 - I(t)$, which is non-increasing, starts from one at time 0, and remains constant until time τ, when it drops down to zero. To obtain a martingale we can multiply $1 - I(t)$ by a suitable increasing function equal to one at $t = 0$.

Proposition 3.17
The process

$$L(t) = (1 - I(t))e^{\Gamma(t)}$$

is a martingale.

Proof Let $0 \le s < t$. Then, using Proposition 2.27, we have

$$\mathbb{E}(L(t)|\mathcal{I}_s) = \mathbf{1}_{\{\tau \le s\}}\mathbb{E}(L(t)|\sigma(\tau)) + \mathbf{1}_{\{s < \tau\}} \frac{\mathbb{E}(L(t)\mathbf{1}_{\{s < \tau\}})}{Q(s < \tau)}$$

$$= e^{\Gamma(t)}\mathbf{1}_{\{\tau \le s\}}\mathbf{1}_{\{t < \tau\}} + e^{\Gamma(t)}\mathbf{1}_{\{s < \tau\}} \frac{\mathbb{E}(\mathbf{1}_{\{t < \tau\}}\mathbf{1}_{\{s < \tau\}})}{Q(s < \tau)}$$

$$= 0 + \mathbf{1}_{\{s < \tau\}}e^{\Gamma(s)}e^{\Gamma(t)}\mathbb{E}(\mathbf{1}_{\{t < \tau\}})$$

$$= L(s)e^{\Gamma(t)}G(t)$$

$$= L(s).$$

$\square$

Exercise 3.12 From the fact that the process $(1 - I(t))e^{\Gamma(t)}$ is a martingale, deduce that

$$\mathbb{E}(\mathbf{1}_{\{T<\tau\}}|\mathcal{I}_t) = \mathbf{1}_{\{t<\tau\}}e^{-\int_t^T \gamma(s)ds}.$$

3.5 Martingale properties of integrals

We are moving towards the next major goal, a characterisation of the process $D(t, T)$ by means of an integral equation. Randomness comes from τ, so it is natural to consider an integral with respect to the indicator process $I(t)$. However, a better choice is to integrate with respect to the compensated indicator $M(t) = I(t) - \Gamma(t \wedge \tau)$. This is because the desired martingale property is easier to obtain when we integrate with respect to a martingale.

Because M is the difference of two processes with non-decreasing paths, we can understand the integral of a process X with respect to M as the Lebesgue–Stieltjes integral $\int_0^t X(s)dM(s)$ defined separately for each path, as long as this integral exists for Q-almost every path. We then say that $\int_0^t X(s)dM(s)$ is a **pathwise** Lebesgue–Stieltjes integral.

By the definition and properties of Lebesgue–Stieltjes integrals (see Appendix A.1), we have

$$\int_0^t X(s)dM(s) = \int_0^t X(s)dI(s) - \int_0^t X(s)d\Gamma(s \wedge \tau)$$

$$= \mathbf{1}_{\{\tau \le t\}}X(\tau) - \int_0^{t \wedge \tau} X(s)\gamma(s)ds. \qquad (3.4)$$

Exercise 3.13 Show that $\int_0^t X(u)dM(u)$ is a martingale for $X(t) = \mathbf{1}_{[0,c]}(t)$, where $c \ge 0$ is a constant.

We need to identify a class of processes X such that $\int_0^t X(s)dM(s)$ is a martingale, and begin with a simple but instructive example.

Example 3.18

For $X(t) = \mathbf{1}_{[0,\tau)}(t) = \mathbf{1}_{\{t<\tau\}}$, which is right-continuous at τ, the integral

$$
\begin{aligned}
\int_0^t X(s)dM(s) &= \mathbf{1}_{\{\tau\leq t\}}X(\tau) - \int_0^{t\wedge\tau} X(s)\gamma(s)ds \\
&= \mathbf{1}_{\{\tau\leq t\}}\mathbf{1}_{\{\tau<\tau\}} - \int_0^{t\wedge\tau} \mathbf{1}_{\{s<\tau\}}\gamma(s)ds \\
&= -\int_0^{t\wedge\tau} \gamma(s)ds \\
&= -\Gamma(t\wedge\tau)
\end{aligned}
$$

is not a martingale since the necessary condition of constant expectation is violated. Indeed, by Exercise 3.10,

$$
\mathbb{E}\left(\int_0^t X(s)dM(s)\right) = -\mathbb{E}\left(\Gamma(t\wedge\tau)\right) = -F(t).
$$

However, for the left-continuous process $X(t) = \mathbf{1}_{[0,\tau]}(t) = \mathbf{1}_{\{t\leq\tau\}}$ the integral

$$
\begin{aligned}
\int_0^t X(s)dM(s) &= \mathbf{1}_{\{\tau\leq t\}}X(\tau) - \int_0^{t\wedge\tau} X(s)\gamma(s)ds \\
&= \mathbf{1}_{\{\tau\leq t\}}\mathbf{1}_{\{\tau\leq\tau\}} - \int_0^{t\wedge\tau} \mathbf{1}_{\{s\leq\tau\}}\gamma(s)ds \\
&= I(t) - \int_0^{t\wedge\tau} \gamma(s)ds \\
&= I(t) - \Gamma(t\wedge\tau)
\end{aligned}
$$

is a martingale by Theorem 3.15.

Motivated by Example 3.18, we shall take X to be a process with left-continuous paths, on top of the standard condition that X should be adapted to the filtration $(\mathcal{I}_t)_{t\geq 0}$. Before considering this case, we prove that the integral $\int_0^t X(u)dM(u)$ is a martingale when X is deterministic.

Lemma 3.19

If $X(t)$ is a deterministic function defined for $t \in [0,\infty)$ and such that $X(\tau)$ is integrable and the Lebesgue–Stieltjes integral $\int_0^t X(u)dM(u)$ exists for all $t \geq 0$, then $\int_0^t X(u)dM(u)$ is a martingale with respect to the filtration $(\mathcal{I}_t)_{t\geq 0}$.

Proof First of all, we need to show that $\int_0^t X(u)dM(u)$ is an integrable random variable. According to (3.4),

$$\int_0^t X(u)dM(u) = \mathbf{1}_{\{\tau \leq t\}} X(\tau) - \int_0^{t \wedge \tau} X(u)\gamma(u)du. \qquad (3.5)$$

Because $X(\tau)$ is integrable, so is the term $\mathbf{1}_{\{\tau \leq t\}} X(\tau)$. To show that the other term on the right-hand side is also integrable we observe that

$$\mathbb{E}\left|\int_0^{t \wedge \tau} X(u)\gamma(u)du\right| \leq \mathbb{E}\left(\int_0^{t \wedge \tau} |X(u)|\,\gamma(u)du\right)$$

$$= \int_0^\infty \left(\int_0^{t \wedge v} |X(u)|\,\gamma(u)du\right)f(v)dv.$$

By changing the order of integration, we obtain

$$\mathbb{E}\left|\int_0^{t \wedge \tau} X(u)\gamma(u)du\right| \leq \int_0^t |X(u)|\,\gamma(u)\left(\int_u^\infty f(v)dv\right)du$$

$$= \int_0^t |X(u)|\,\gamma(u)G(u)du$$

$$= \int_0^t |X(u)|\,f(u)du$$

$$= \mathbb{E}\left(\mathbf{1}_{\{\tau \leq t\}} |X(\tau)|\right)$$

$$\leq \mathbb{E}|X(\tau)| < \infty$$

since $\gamma(u)G(u) = f(u)$. This shows that both terms on the right-hand side of (3.5) are integrable, hence $\int_0^t X(u)dM(u)$ is integrable.

Our next goal is to show that for any $0 \leq s < t$

$$\mathbb{E}\left(\int_0^t X(u)dM(u) \,\Big|\, \mathcal{I}_s\right) = \int_0^s X(u)dM(u). \qquad (3.6)$$

The right-hand side is given by (3.4), while for the left-hand side we use Proposition 2.27, which gives

$$\mathbb{E}\left(\int_0^t X(u)dM(u) \,\Big|\, \mathcal{I}_s\right)$$

$$= \mathbf{1}_{\{\tau \leq s\}}\mathbb{E}\left(\int_0^t X(u)dM(u) \,\Big|\, \sigma(\tau)\right) + \mathbf{1}_{\{s < \tau\}}e^{\Gamma(s)}\mathbb{E}\left(\mathbf{1}_{\{s < \tau\}}\int_0^t X(u)dM(u)\right).$$

$$(3.7)$$

Insert (3.4), and the first term on the right-hand side takes the form

$$\mathbf{1}_{\{\tau \le s\}} \mathbb{E}\left(\int_0^t X(u)dM(u) \Big| \sigma(\tau)) \right)$$

$$= \mathbf{1}_{\{\tau \le s\}} \mathbb{E}\left(\mathbf{1}_{\{\tau \le t\}} X(\tau) - \int_0^{t \wedge \tau} X(u)\gamma(u)du \Big| \sigma(\tau) \right)$$

$$= \mathbf{1}_{\{\tau \le s\}} \left(\mathbf{1}_{\{\tau \le t\}} X(\tau) - \mathbf{1}_{\{\tau \le t\}} \int_0^{\tau} X(u)\gamma(u)du - \mathbf{1}_{\{t < \tau\}} \int_0^t X(u)\gamma(u)du \right)$$

$$= \mathbf{1}_{\{\tau \le s\}} X(\tau) - \mathbf{1}_{\{\tau \le s\}} \int_0^{\tau} X(u)\gamma(u)du.$$

Next we compute the expectation

$$\mathbb{E}\left(\mathbf{1}_{\{s < \tau\}} \int_0^t X(u)dM(u) \right)$$

$$= \mathbb{E}\left(\mathbf{1}_{\{s < \tau\}} \left(\mathbf{1}_{\{\tau \le t\}} X(\tau) - \int_0^{t \wedge \tau} X(u)\gamma(u)du \right) \right)$$

$$= \mathbb{E}\left(\mathbf{1}_{\{s < \tau \le t\}} X(\tau) \right) - \mathbb{E}\left(\mathbf{1}_{\{s < \tau\}} \int_0^t \mathbf{1}_{\{u < \tau\}} X(u)\gamma(u)du \right)$$

$$= \int_s^t X(u)f(u)du - \mathbb{E}\left(\int_0^t \mathbf{1}_{\{u \vee s < \tau\}} X(u)\gamma(u)du \right)$$

$$= \int_s^t X(u)f(u)du - \int_0^t G(u \vee s)X(u)\gamma(u)du$$

$$= \int_s^t X(u)f(u)du - G(s) \int_0^s X(u)\gamma(u)du - \int_s^t G(u)X(u)\gamma(u)du$$

$$= -e^{-\Gamma(s)} \int_0^s X(u)\gamma(u)du.$$

Substituting this into (3.7), we obtain

$$\mathbb{E}\left(\int_0^t X(u)dM(u) \Big| \mathcal{I}_s \right)$$

$$= \mathbf{1}_{\{\tau \le s\}} \mathbb{E}\left(\int_0^t X(u)dM(u) \Big| \sigma(\tau) \right) + \mathbf{1}_{\{s < \tau\}} e^{\Gamma(s)} \mathbb{E}\left(\mathbf{1}_{\{s < \tau\}} \int_0^t X(u)dM(u) \right)$$

$$= \mathbf{1}_{\{\tau \le s\}} X(\tau) - \mathbf{1}_{\{\tau \le s\}} \int_0^{\tau} X(u)\gamma(u)du - \mathbf{1}_{\{s < \tau\}} \int_0^s X(u)\gamma(u)du$$

$$= \int_0^s X(u)dM(u),$$

which shows that $\int_0^t X(u)dM(u)$ is a martingale. $\qquad\square$

Exercise 3.14 Take any $0 \le a < b$ and any $\mathcal{I}_a$-measurable random variable Z, and let $X(t) = Z\mathbf{1}_{(a,b]}(t)$ for $t \ge 0$. Show that there is a deterministic function $Y(t)$ such that

$$\int_0^t X(u)dM(u) = \int_0^t Y(u)dM(u),$$

and deduce that $\int_0^t X(u)dM(u)$ is a martingale with respect to the filtration $(\mathcal{I}_t)_{t\ge 0}$.

We are ready to establish the martingale property of $\int_0^t X(u)dM(u)$ in the case when X is an adapted process with left-continuous paths.

Theorem 3.20
If X is a stochastic process with left-continuous paths, adapted to the filtration $(\mathcal{I}_t)_{t\ge 0}$ and such that $X(\tau)$ is integrable and the Lebesgue–Stieltjes integral $\int_0^t X(u)dM(u)$ exists for all $t \ge 0$, then $\int_0^t X(u)dM(u)$ is an $(\mathcal{I}_t)_{t\ge 0}$-martingale.

Proof For any $t \ge 0$, since $X(t)$ is $\mathcal{I}_t$-measurable, by Proposition 2.22 it can be written as

$$X(t) = \eta_t(\tau)\mathbf{1}_{\{\tau \le t\}} + c_t\mathbf{1}_{\{t < \tau\}}$$

for some Borel function $\eta_t : \mathbb{R} \to \mathbb{R}$ and a deterministic number c_t. We are going to show that c_t is left-continuous as a function of $t \in [0, \infty)$, and such that c_τ is integrable and

$$\int_0^t c_u dM(u) = \int_0^t X(u)dM(u).$$

Given that X has left-continuous paths and $\{t < \tau\} \subset \{s < \tau\}$ when $0 \le s < t$, on the event $\{t < \tau\}$ we have

$$\lim_{s \nearrow t} c_s = \lim_{s \nearrow t} X(s) = X(t) = c_t.$$

Because c_t is deterministic and $\{t < \tau\}$ has positive probability for any t, it follows that c_t is indeed left-continuous as a function of t. Next, since X has left-continuous paths and c_t is left-continuous as a function of t,

$$X(\tau) = \lim_{s \nearrow \tau} X(s) = \lim_{s \nearrow \tau} c_s = c_\tau.$$

It follows that c_τ is integrable since $X(\tau)$ is. Now consider

$$\int_0^t c_u dM(u) = \mathbf{1}_{\{\tau \le t\}} c_\tau - \int_0^{t \wedge \tau} c_u \gamma(u) du$$

$$= \mathbf{1}_{\{\tau \le t\}} X(\tau) - \int_0^{t \wedge \tau} X(u) \gamma(u) du$$

$$= \int_0^t X(u) dM(u).$$

By Lemma 3.19, we can conclude that $\int_0^t c_u dM(u)$, and so $\int_0^t X(u) dM(u)$ is a martingale with respect to the filtration $(\mathcal{I}_t)_{t \ge 0}$. □

3.6 Bond price dynamics

We would like to characterise defaultable bond prices by means of an integral equation. This will make it possible to define the self-financing property for general strategies, not necessarily simple (i.e. piecewise constant) ones as in Definition 2.33.

First, we obtain an equation for the bond price by performing a crude heuristic computation involving differentials. Namely, given that $D(t, T) = (1 - I(t))\hat{D}(t, T)$, where $\hat{D}(t, T) = e^{-r(T-t)} e^{-(\Gamma(T) - \Gamma(t))}$, we can write

$$dD = d(\hat{D}(1 - I))$$

$$= (1 - I) d\hat{D} - \hat{D} dI$$

$$= (1 - I) r \hat{D} dt + (1 - I) \gamma \hat{D} dt - \hat{D} dI$$

$$= rD dt - [\hat{D} dI - (1 - I) \hat{D} \gamma dt]$$

$$= rD dt - [\hat{D} dI - \mathbf{1}_{\{\tau \le t\}} \gamma \hat{D} dt]$$

$$= rD dt - \hat{D} dM.$$

This helps us to formulate the following rigorous version.

Theorem 3.21
The defaultable bond price process $D(t, T)$ satisfies the integral equation

$$D(t, T) - D(0, T) = \int_0^t rD(u, T) du - \int_0^t \hat{D}(u, T) dM(u) \qquad (3.8)$$

for all $t \ge 0$, where $M(t)$ is the compensated indicator process and $\hat{D}(t, T)$ is the pre-default value.

Proof Using formula (3.4) and the fact that $D(t, T) = \mathbf{1}_{\{t<\tau\}}\hat{D}(t, T)$, we can write (3.8) as

$$D(t, T) - \hat{D}(0, T) = \int_0^{t\wedge\tau} r\hat{D}(u, T)du - \mathbf{1}_{\{\tau\le t\}}\hat{D}(\tau, T) + \int_0^{t\wedge\tau} \hat{D}(u, T)\gamma(u)du.$$

On $\{t < \tau\}$ we have $D(t, T) = \hat{D}(t, T)$, and (3.8) becomes

$$\hat{D}(t, T) - \hat{D}(0, T) = \int_0^t r\hat{D}(u, T)du + \int_0^t \hat{D}(u, T)\gamma(u)du,$$

which holds according to equation (3.3). On the other hand, on $\{\tau \le t\}$ we have $D(t, T) = 0$, and (3.8) becomes

$$-\hat{D}(0, T) = \int_0^\tau r\hat{D}(u, T)du - \hat{D}(\tau, T) + \int_0^\tau \hat{D}(u, T)\gamma(u)du.$$

This is equation (3.3) once again.   □

Remark 3.22
Equation (3.8) for $D(t, T)$ would be more elegant if it involved D only rather than both D and $\hat{D}$. For $t < \tau$ we have $D(t, T) = \hat{D}(t, T)$. For $\tau < t$ we do not have this equality, but this is irrelevant since $\int_0^t \hat{D}(u, T)dM(u)$ does not depend on the values of $\hat{D}(t, T)$ for $\tau < t$. At time τ we have

$$\hat{D}(\tau, T) = \lim_{t\nearrow\tau} D(t, T) = D(t_-, T).$$

Hence, we can write (3.8) as

$$D(t, T) - D(0, T) = \int_0^t rD(u, T)du - \int_0^t D(u_-, T)dM(u).$$

Proposition 3.23
The discounted bond prices $\tilde{D}(t, T) = e^{-rt}D(t, T)$ satisfy

$$\tilde{D}(t, T) - \tilde{D}(0, T) = -\int_0^t \tilde{D}(u_-, T)dM(u).$$

Proof Using the properties of Lebesgue–Stieltjes integrals, namely the integration-by-parts formula, Theorem A.9, followed by Theorems A.10

and A.11 in Appendix A.1, we obtain

$$
\begin{aligned}
e^{-rt}D(t,T) - D(0,T) &= \int_0^t e^{-ru}dD(u,T) + \int_0^t D(u_-,T)d(e^{-ru}) \\
&= \int_0^t re^{-ru}D(u,T)du - \int_0^t e^{-ru}D(u_-,T)dM(u) \\
&\quad - \int_0^t re^{-tu}D(u_-,T)du \\
&= -\int_0^t \tilde{D}(u_-,T)dM(u),
\end{aligned}
$$

given that $D(u,T) = D(u_-,T)$ for each $u \in [0,t]$ except for $u = \tau$. $\quad\square$

Remark 3.24

According to Theorem 3.20, $\int_0^t \tilde{D}(u_-,T)dM(u)$ is a martingale. This is consistent with the discounted bond price process $\tilde{D}(t,T) = e^{-rt}D(t,T)$ being a martingale under the risk-neutral probability Q (Assumption 3.1).

4

Security pricing with hazard function

4.1   Martingale representation theorem

Financial securities can be defined in terms of the associated cash flows. The simplest case is when there is a single payoff X, an $\mathcal{I}_T$-measurable random variable, due at some fixed time T. Just as in the classical theories, a step towards replicating the payoff will be to represent the martingale $\mathbb{E}(X|\mathcal{I}_t)$ as an integral with respect to the driving process $I(t)$ or, more conveniently, with respect to the compensated process $M(t) = I(t) - \Gamma(t \wedge \tau)$.

Since $\mathcal{I}_T \subset \sigma(\tau)$, it follows that X is $\sigma(\tau)$-measurable, so we know that there is a Borel function h such that $X = h(\tau)$. This motivates the following result.

Theorem 4.1
For any Borel function h such that $h(\tau)$ is integrable with respect to Q, the martingale $H(t) = \mathbb{E}(h(\tau)|\mathcal{I}_t)$ satisfies

$$H(t) = \mathbb{E}(h(\tau)) + \int_0^t (h(s) - J(s))dM(s), \qquad (4.1)$$

where

$$J(t) = \mathbb{E}(h(\tau)|t < \tau)$$

and where $M(t) = I(t) - \Gamma(t \wedge \tau)$ is the compensated indicator process.

94

Proof We begin with some alternative expressions for J, namely

$$J(t) = \mathbb{E}(h(\tau)|t < \tau) = \frac{\mathbb{E}(1_{\{t<\tau\}}h(\tau))}{Q(t < \tau)}$$

$$= e^{\Gamma(t)}\mathbb{E}(1_{\{t<\tau\}}h(\tau)) = e^{\Gamma(t)}\int_t^\infty h(u)f(u)du.$$

By using Proposition 2.27, we can write $H(t)$ as

$$H(t) = \mathbb{E}(h(\tau)|\mathcal{I}_t)$$

$$= 1_{\{\tau \le t\}}\mathbb{E}(h(\tau)|\sigma(\tau)) + 1_{\{t<\tau\}}\frac{\mathbb{E}(1_{\{t<\tau\}}h(\tau))}{Q(t < \tau)}$$

$$= 1_{\{\tau \le t\}}h(\tau) + 1_{\{t<\tau\}}e^{\Gamma(t)}\mathbb{E}(1_{\{t<\tau\}}h(\tau))$$

$$= 1_{\{\tau \le t\}}h(\tau) + 1_{\{t<\tau\}}J(t).$$

By formula (3.4), the right-hand side of (4.1) can be written as

$$\mathbb{E}(h(\tau)) + \int_0^t (h(s) - J(s))dM(s)$$

$$= \mathbb{E}(h(\tau)) + 1_{\{\tau \le t\}}(h(\tau) - J(\tau)) - \int_0^{t\wedge\tau} (h(u) - J(u))\gamma(u)du.$$

On comparing the above expressions, our task is reduced to proving that

$$1_{\{t<\tau\}}J(t) + 1_{\{\tau \le t\}}J(\tau) = \mathbb{E}(h(\tau)) - \int_0^{t\wedge\tau} h(s)\gamma(s)ds + \int_0^{t\wedge\tau} J(s)\gamma(s)ds. \quad (4.2)$$

We evaluate the last integral as

$$\int_0^{t\wedge\tau} J(s)\gamma(s)ds = \int_0^{t\wedge\tau} e^{\Gamma(s)}\left(\int_s^\infty h(u)f(u)du\right)\gamma(s)ds$$

$$= \int_0^{t\wedge\tau} \left(\int_s^\infty h(u)f(u)du\right)d(e^{\Gamma(s)})$$

$$= e^{\Gamma(t\wedge\tau)}\int_{t\wedge\tau}^\infty h(u)f(u)du - e^{\Gamma(0)}\int_0^\infty h(u)f(u)du$$

$$+ \int_0^{t\wedge\tau} e^{\Gamma(s)}h(s)f(s)ds$$

using the integration-by-parts formula. We have $\Gamma(0) = 0$ and

$$\int_0^\infty h(u)f(u)du = \mathbb{E}(h(\tau)),$$

so the task reduces to showing that

$$\mathbf{1}_{\{t<\tau\}}J(t) + \mathbf{1}_{\{\tau\le t\}}J(\tau) = -\int_0^{t\wedge\tau} h(s)\gamma(s)ds + e^{\Gamma(t\wedge\tau)}\int_{t\wedge\tau}^\infty h(u)f(u)du$$
$$+ \int_0^{t\wedge\tau} e^{\Gamma(s)}h(s)f(s)ds.$$

Bearing in mind that

$$\mathbf{1}_{\{t<\tau\}}J(t) + \mathbf{1}_{\{\tau\le t\}}J(\tau) = J(t\wedge\tau) = e^{\Gamma(t\wedge\tau)}\int_{t\wedge\tau}^\infty h(u)f(u)du,$$

our task (4.2) reduces to noticing that

$$\int_0^{t\wedge\tau} h(s)\gamma(s)ds = \int_0^{t\wedge\tau} e^{\Gamma(s)}h(s)f(s)ds$$

as a result of the relationship $f(s) = e^{-\Gamma(s)}\gamma(s)$. $\qquad\qquad\square$

Exercise 4.1 According to Assumption 3.1, the discounted default-able bond price process $H(t) = e^{-rt}D(t,T)$ is an $(\mathcal{I}_t)_{t\ge0}$-martingale under Q. Find the corresponding functions $h(t)$ and $J(t)$ and verify equality (4.1).

Exercise 4.2 Suppose that an $(\mathcal{I}_t)_{t\ge0}$-martingale X can be represented as

$$X(t) = X(0) + \int_0^t \phi(s)dM(s)$$

for all $t \ge 0$, where ϕ is a left-continuous deterministic function. Show that such a function ϕ is unique.

4.2 Pricing defaultable securities

With the dynamics of a defaultable bond in hand (see Section 3.6), we are ready to formulate the self-financing condition for more general trading strategies than in Section 2.4. The strategies will no longer need to be piecewise constant.

Definition 4.2

A trading **strategy** in the market model consisting of the non-defaultable bond B and defaultable bond D is a pair $\varphi = (\varphi_B, \varphi_D)$ of $(\mathcal{I}_t)_{t \geq 0}$-adapted processes having left-continuous paths. The **value** of such a strategy φ at time $t \in [0, T]$ is given by

$$V_\varphi(t) = \varphi_B(t)B(t, T) + \varphi_D(t)D(t, T).$$

The strategy is said to be **self-financing** if for all $t \in [0, T]$

$$V_\varphi(t) = V_\varphi(0) + \int_0^t \varphi_B(u)dB(u, T) + \int_0^t \varphi_D(u)dD(u, T). \qquad (4.3)$$

The integrals in this definition and in what follows are understood as Lebesgue–Stieltjes integrals; see Appendix A.1.

Exercise 4.3 Show that the discounted values $\tilde{V}_\varphi(t) = e^{-rt}V_\varphi(t)$ of a self-financing strategy φ satisfy

$$\tilde{V}_\varphi(t) = V_\varphi(0) + \int_0^t \varphi_D(u)d\tilde{D}(u, T)$$

for all $t \in [0, T]$.

As a consequence of Exercise 4.3, given $V(0)$ and $\varphi_D(t)$, we can find $\varphi_B(t)$ such that the strategy $\varphi = (\varphi_B, \varphi_D)$ is self-financing, namely

$$\varphi_B(t) = \frac{1}{e^{-rt}B(t, T)}\left(V_\varphi(0) + \int_0^t \varphi_D(u)d\tilde{D}(u, T) - \varphi_D(t)\tilde{D}(t, T) \right). \qquad (4.4)$$

Moreover, it follows from Exercise 4.3 that the value process $\tilde{V}_\varphi$ will be an $(\mathcal{I}_t)_{t \geq 0}$-martingale if $\varphi_D(\tau)$ is integrable. Indeed, in this case, the random variable $\varphi_D(\tau)\tilde{D}(\tau_-, T)$ will also be integrable. In addition, the process $\varphi_D(t)\tilde{D}(t-, T)$ will have left-continuous paths, so the integral

$$\int_0^t \varphi_D(u)d\tilde{D}(u, T) = -\int_0^t \varphi_D(u)\tilde{D}(u_-, T)dM(u)$$

will be an $(\mathcal{I}_t)_{t \geq 0}$-martingale by Theorem 3.20. Observe that the last equality is a consequence of Theorem A.11.

Definition 4.3

We say that a self-financing strategy $\varphi = (\varphi_B, \varphi_D)$ is **admissible** if the discounted value process $\tilde{V}_\varphi$ is an $(\mathcal{I}_t)_{t \geq 0}$-martingale under Q.

Example 4.4

Consider the following strategy with right-continuous rather than left-continuous paths. At time 0 we buy one defaultable bond and short the default-free bond, so that $V_\varphi(0) = 0$. Then we keep $\varphi_D(t) = 1$ on $\{t < \tau\}$ and put $\varphi_D(t) = 0$ on $\{\tau \le t\}$. That is, we take

$$\varphi_D(t) = \mathbf{1}_{\{t<\tau\}}.$$

We also take the non-defaultable bond position as given by (4.4). Hence, by using Proposition 3.23 and substituting

$$\tilde{D}(t,T) = \mathbf{1}_{\{t<\tau\}} e^{-rT} e^{-\Gamma(T)} e^{\Gamma(t)},$$

we can express the integral in (4.4) as

$$
\begin{aligned}
\int_0^t \varphi_D(u) d\tilde{D}(u,T) &= -\int_0^t \varphi_D(u)\tilde{D}(u_-,T)dM(u) \\
&= -\mathbf{1}_{\{\tau\le t\}}\varphi_D(\tau)\tilde{D}(\tau_-,T) + \int_0^{t\wedge\tau} \varphi_D(u)\tilde{D}(u_-,T)\gamma(u)du \\
&= e^{-rT} e^{-\Gamma(T)} \int_0^{t\wedge\tau} e^{\Gamma(u)}\gamma(u)du \\
&= e^{-rT} e^{-\Gamma(T)} \left(e^{\Gamma(t\wedge\tau)} - 1 \right).
\end{aligned}
$$

It follows that

$$
\begin{aligned}
\varphi_B(t) &= \frac{1}{e^{-rt}B(t,T)} \left(V_\varphi(0) + \int_0^t \varphi_D(u)d\tilde{D}(u,T) - \varphi_D(t)\tilde{D}(t,T) \right) \\
&= e^{-\Gamma(T)} \left(e^{\Gamma(t\wedge\tau)} - 1 - \mathbf{1}_{\{t<\tau\}} e^{\Gamma(t)} \right) \\
&= e^{-\Gamma(T)} \left(\mathbf{1}_{\{\tau\le t\}} e^{\Gamma(\tau)} - 1 \right).
\end{aligned}
$$

This shows that on $\{\tau \le T\}$ the default-free bond position jumps up by $e^{-\Gamma(T)} e^{\Gamma(\tau)}$ at time τ. It implies an injection of cash into the strategy, as the jump cannot be financed by selling the defaultable bond, which becomes worthless, $D(\tau,T) = 0$, at time τ. But formula (4.4) for $\varphi_B(t)$ is a consequence of the self-financing condition (4.3). This forces us to conclude that the above strategy with right-continuous paths is incompatible with the self-financing condition (4.3).

Also observe that the discounted value process

$$
\begin{aligned}
\tilde{V}_\varphi(t) &= e^{-rt}(\varphi_B(t)B(t,T) + \varphi_D(t)D(t,T)) \\
&= e^{-rT} e^{-\Gamma(T)}(e^{\Gamma(t\wedge\tau)} - 1)
\end{aligned}
$$

is not a martingale under Q because the expectation $\mathbb{E}(\tilde{V}_\varphi(T)) = e^{-rT}e^{-\Gamma(T)}\Gamma(T)$, computed in Exercise 4.4, is not equal to $\tilde{V}_\varphi(0) = 0$. This means that $\varphi = (\varphi_B, \varphi_D)$ cannot be an admissible strategy.

Exercise 4.4 Compute the expectation $\mathbb{E}(\tilde{V}_\varphi(T))$ and show that this is not equal to zero.

Example 4.5

To avoid the problems highlighted in Example 4.4, our strategies need to be left-continuous. Indeed, if we take

$$\varphi_D(t) = \mathbf{1}_{\{t \leq \tau\}},$$

which is left-continuous, then

$$\int_0^t \varphi_D(u)d\tilde{D}(u, T) = -\int_0^t \varphi_D(u)\tilde{D}(u_-, T)dM(u)$$

$$= -\mathbf{1}_{\{\tau \leq t\}}\varphi_D(\tau)\tilde{D}(\tau_-, T) + \int_0^{t \wedge \tau} \varphi_D(u)\tilde{D}(u_-, T)\gamma(u)du$$

$$= e^{-rT}e^{-\Gamma(T)}\left(-\mathbf{1}_{\{\tau \leq t\}}e^{\Gamma(\tau)} + \int_0^{t \wedge \tau} e^{\Gamma(u)}\gamma(u)du\right)$$

$$= e^{-rT}e^{-\Gamma(T)}\left(-\mathbf{1}_{\{\tau \leq t\}}e^{\Gamma(\tau)} + e^{\Gamma(t \wedge \tau)} - 1\right),$$

and (4.4) with $V_\varphi(0) = 0$ gives

$$\varphi_B(t) = \frac{1}{e^{-rt}B(t, T)}\left(V_\varphi(0) + \int_0^t \varphi_D(u)d\tilde{D}(u, T) - \varphi_D(t)\tilde{D}(t, T)\right)$$

$$= e^{-\Gamma(T)}\left(-\mathbf{1}_{\{\tau \leq t\}}e^{\Gamma(\tau)} + e^{\Gamma(t \wedge \tau)} - 1 - \mathbf{1}_{\{t < \tau\}}e^{\Gamma(t)}\right)$$

$$= -e^{-\Gamma(T)}.$$

The default-free position is constant, there is no jump at time τ, and no cash injection, in full agreement with the idea of self-financing.

In this case, since $\varphi_D(t)$ is left-continuous and $\varphi_D(\tau)$ is integrable, it follows that the process $\varphi_D(t)\tilde{D}(t_-, T)$ has left-continuous paths and

$\varphi_D(\tau)\tilde{D}(\tau_-, T)$ is integrable. By Theorem 3.20, the discounted value process

$$\tilde{V}_\varphi(t) = \int_0^t \varphi_D(u)d\tilde{D}(u, T) = -\int_0^t \varphi_D(u)\tilde{D}(u_-, T)dM(u)$$

is a martingale under Q. Hence, $\varphi = (\varphi_B, \varphi_D)$ is an admissible strategy.

Definition 4.6
An **arbitrage strategy** is a self-financing strategy φ such that $V_\varphi(0) = 0$, $V_\varphi(T) \geq 0$ almost surely, and $V_\varphi(T) > 0$ with positive probability.

Exercise 4.5 Show that the existence of a risk-neutral probability implies that there is no admissible arbitrage strategy.

We are going to consider a contingent claim of European type with exercise time $T > 0$ whose payoff H is an $\mathcal{I}_T$-measurable random variable. Our aim is to determine the process of prices $H(t)$ of such a contingent claim for $t \in [0, T]$.

Definition 4.7
We say that a trading strategy $\varphi = (\varphi_B, \varphi_D)$ in the market consisting of the non-defaultable bond B and defaultable bond D **replicates** the payoff H if $V_\varphi(T) = H$.

We can also consider trading strategies in the extended market consisting of the default-free bond B, defaultable bond D, and contingent claim H. Because we do not yet know the dynamics of the contingent claim price process $H(t)$, we only consider strategies with a static position in H, so that the self-financing condition for a strategy in the extended market B, D, H reduces to that in the B, D market.

Definition 4.8
An **arbitrage opportunity** in the extended market model consisting of the assets B, D and contingent claim H is a self-financing strategy $\psi = (\psi_B, \psi_D)$ in the B, D market and a static position $z = 1$ or -1 in the contingent claim H such that $V_\psi(0) + zH(0) = 0$, $V_\psi(T) + zH(T) \geq 0$, and $V_\psi(T) + zH(T) > 0$ with positive probability.

Theorem 4.9

Suppose that $\varphi = (\varphi_B, \varphi_D)$ is a self-financing strategy in the B, D market that replicates a contingent claim with $\mathcal{I}_T$-measurable payoff H and exercise time T. If there are no arbitrage opportunities in the extended market B, D, H, then for each time $t \in [0, T]$ the price of the contingent claim must be equal to the value of the replicating strategy,

$$H(t) = V_\varphi(t).$$

Proof It is enough to verify this equality just for $t = 0$. This is because any time instant prior to T can be chosen as the initial time 0 without loss of generality.

Suppose that the equality is violated at time 0, say $H(0) > V_\varphi(0)$. In this case, to construct an arbitrage opportunity we can enter a long position in the replicating strategy, a short position in the contingent claim, and invest the difference $H(0) - V_\varphi(0) > 0$ in the non-defaultable bond. To be precise, we can follow the self-financing strategy in the B, D market defined by

$$(\psi_B(t), \psi_D(t)) = (\varphi_B(t) + B(0, T)^{-1}(H(0) - V_\varphi(0)), \varphi_D(t))$$

for each $t \in [0, T]$, and take the position $z = -1$ in the contingent claim H. Then

$$V_\psi(0) + zH(0) = V_\varphi(0) + (H(0) - V_\varphi(0)) - H(0) = 0$$

and

$$\begin{aligned} V_\psi(T) + zH(T) &= V_\varphi(T) + B(0, T)^{-1}(H(0) - V_\varphi(0)) - H \\ &= B(0, T)^{-1}(H(0) - V_\varphi(0)) > 0 \end{aligned}$$

since the replicating strategy satisfies $V_\varphi(T) = H$. This is indeed an arbitrage opportunity in the extended market B, D, H. On the other hand, when $H(0) > V_\varphi(0)$, the opposite positions also produce an arbitrage opportunity. It follows that $H(0) = V_\varphi(0)$ if no arbitrage opportunities exist in the extended market. $\square$

Next, we are going to show that any $\mathcal{I}_T$-measurable payoff H can be replicated. Such a random variable is of the form $H = h(\tau)$ for some Borel function h since $\mathcal{I}_T \subset \sigma(\tau)$, so H is $\sigma(\tau)$-measurable.

Example 4.10

As a warm-up, we consider a defaultable bond that pays a constant positive amount $0 < \delta < 1$ at maturity T when a default occurs before or at time T, or pays 1 when there is no default up to and including time T, that is, a defaultable bond with constant recovery. The payoff can be written as

$$H = \mathbf{1}_{\{T<\tau\}} + \delta\mathbf{1}_{\{\tau\leq T\}}$$
$$= \delta + (1-\delta)\mathbf{1}_{\{T<\tau\}}$$
$$= \delta + (1-\delta)D(T,T).$$

The price of this bond is denoted by $D_\delta(t, T)$. In particular,

$$D_\delta(T,T) = \delta + (1-\delta)D(T,T).$$

Replication can be achieved by taking

$$\varphi_B(t) = \delta,$$
$$\varphi_D(t) = 1 - \delta$$

for each $t \in [0, T]$. It is a self-financing strategy:

$$\delta B(t,T) + (1-\delta)D(t,T) = \delta B(0,T) + (1-\delta)D(0,T)$$
$$+ \delta \int_0^t dB(t,T) + (1-\delta) \int_0^t dD(t,T).$$

This immediately gives

$$D_\delta(t,T) = \delta B(t,T) + (1-\delta)D(t,T)$$
$$= \delta e^{-r(T-t)} + (1-\delta)\mathbf{1}_{\{t<\tau\}}e^{-r(T-t)}e^{-(\Gamma(T)-\Gamma(t))}.$$

Observe that the values

$$V_\varphi(t) = \delta B(t,T) + (1-\delta)D(t,T)$$

are non-negative since $0 < \delta < 1$, and the discounted value process

$$\tilde{V}_\varphi(t) = e^{-rt}V_\varphi(t)$$
$$= \delta e^{-rT} + (1-\delta)e^{-rt}D(t,T)$$

is a martingale under Q since $e^{-rt}D(t,T)$ is. It follows that $\varphi = (\varphi_B, \varphi_D)$ is an admissible strategy.

The formula for $D_\delta(t, T)$ is often obtained in the literature by assuming that the discounted prices $e^{-rt}D_\delta(t, T)$ form a martingale and computing

$$D_\delta(t,T) = e^{-r(T-t)}\mathbb{E}(D_\delta(T,T)|\mathcal{I}_t).$$

However, this martingale property follows from replication, and there is in fact no need to assume it.

Exercise 4.6 Compute $e^{-r(T-t)}\mathbb{E}(D_\delta(T,T)|\mathcal{I}_t)$ directly and compare with the formula obtained for $D_\delta(t,T)$ in Example 4.10.

Exercise 4.7 Show that the zero-recovery bond $D(t,T)$ can be replicated by means of the non-defaultable bond $B(t,T)$ and a positive-recovery bond $D_\delta(t,T)$.

Our next task will be to replicate a European option with payoff $h(\tau)$ and expiry date T by means of an admissible self-financing strategy $\varphi = (\varphi_B, \varphi_D)$. We begin with a heuristic argument to find formulae for the components of such a strategy.

If the strategy is to be admissible, its discounted value process $\tilde{V}_\varphi(t) = e^{-rt}V_\varphi(t)$ should be a martingale. Moreover, if the strategy is to replicate the payoff $h(\tau)$ at time T, then $V_\varphi(T) = h(\tau)$. This implies

$$\mathbb{E}(h(\tau)|\mathcal{I}_t) = \mathbb{E}(V_\varphi(T)|\mathcal{I}_t) = e^{rT}\mathbb{E}(\tilde{V}_\varphi(T)|\mathcal{I}_t) = e^{rT}\tilde{V}_\varphi(t) = e^{r(T-t)}V_\varphi(t).$$

Hence, according to the martingale representation theorem,

$$V_\varphi(t)e^{r(T-t)} = V_\varphi(0)e^{rT} + \int_0^t (h(s) - J(s))dM(s), \qquad (4.5)$$

where $J(t) = \mathbb{E}(h(\tau)|t < \tau)$. We are looking for a self-financing strategy, so

$$V_\varphi(t) = V_\varphi(0) + \int_0^t \varphi_B(s)dB(s,T) + \int_0^t \varphi_D(s)dD(s,T).$$

Integration by parts, the formula $B(t,T) = e^{-r(T-t)}$, and equation (3.8) for $D(t,T)$ then give

$$V_\varphi(t)e^{r(T-t)} = V_\varphi(0)e^{rT} - r\int_0^t e^{r(T-s)}V_\varphi(s)ds + \int_0^t e^{r(T-s)}dV_\varphi(s)$$

$$= V_\varphi(0)e^{rT} - r\int_0^t e^{r(T-s)}V_\varphi(s)ds + \int_0^t e^{r(T-s)}\varphi_B(s)dB(s,T)$$

$$+ \int_0^t e^{r(T-s)}\varphi_D(s)dD(s,T)$$

$$= V_\varphi(0)e^{rT} - r\int_0^t e^{r(T-s)}V_\varphi(s)ds + r\int_0^t e^{r(T-s)}\varphi_B(s)B(s,T)ds$$

$$+ r\int_0^t e^{r(T-s)}\varphi_D(s)D(s,T)ds - \int_0^t e^{r(T-s)}\varphi_D(s)\hat{D}(s,T)dM(s).$$

Inserting the expressions $V_\varphi(s) = \varphi_B(s)B(s,T) + \varphi_D(s)D(s,T)$ and $\hat{D}(s,T) = e^{-r(T-s)}e^{-(\Gamma(T)-\Gamma(s))}$, we end up with

$$V_\varphi(t)e^{r(T-t)} = V_\varphi(0)e^{rT} - \int_0^t \varphi_D(s)e^{-(\Gamma(T)-\Gamma(s))}dM(s).$$

By comparing the right-hand sides of this equality and (4.5), we find a candidate for φ_D, namely

$$\varphi_D(t) = e^{\Gamma(T)-\Gamma(t)}(J(t) - h(t)).$$

We also need an expression for $\varphi_B(t)$. Proposition 2.27 shows that

$$e^{r(T-t)}V_\varphi(t) = \mathbb{E}(h(\tau)|\mathcal{I}_t)$$
$$= \mathbf{1}_{\{\tau \le t\}}\mathbb{E}(h(\tau)|\sigma(\tau)) + \mathbf{1}_{\{t < \tau\}}e^{\Gamma(t)}\mathbb{E}(\mathbf{1}_{\{t < \tau\}}h(\tau))$$
$$= \mathbf{1}_{\{\tau \le t\}}h(\tau) + \mathbf{1}_{\{t < \tau\}}J(t).$$

On the other hand, since $D(t,T) = \mathbf{1}_{\{t < \tau\}}e^{-r(T-t)}e^{-(\Gamma(T)-\Gamma(s))}$ and $B(t,T) = e^{-r(T-t)}$, we have

$$e^{r(T-t)}V_\varphi(t) = e^{r(T-t)}(\varphi_B(t)B(t,T) + \varphi_D(t)D(t,T))$$
$$= \varphi_B(t) + (J(t) - h(t))\mathbf{1}_{\{t < \tau\}}.$$

Comparing the right-hand sides, we obtain

$$\varphi_B(t) = \mathbf{1}_{\{\tau \le t\}}h(\tau) + \mathbf{1}_{\{t < \tau\}}h(t) = h(t \wedge \tau).$$

This heuristic argument also offers some hints as to the assumptions for h. Because $h(\tau) = V_\varphi(T) = \tilde{V}_\varphi(T)e^{rT}$ by replication, and because $\tilde{V}_\varphi$ should be a martingale in order for the strategy to be admissible, we need to assume that $h(\tau)$ is integrable and $\mathcal{I}_T$-measurable. Moreover, to satisfy Definition 4.2, the strategy needs to have left-continuous paths. Looking at the formulae for $\varphi_B(t)$ and $\varphi_D(t)$, we can see that this will be the case when h is a left-continuous function. We are ready to state the following important theorem.

Theorem 4.11

If h is a left-continuous function such that $h(\tau)$ is an integrable and $\mathcal{I}_T$-measurable random variable, then there is an admissible self-financing strategy $\varphi = (\varphi_B, \varphi_D)$ such that

$$V_\varphi(t) = e^{-r(T-t)}\mathbb{E}(h(\tau)|\mathcal{I}_t).$$

The components of the strategy are

$$\varphi_B(t) = h(t \wedge \tau),$$
$$\varphi_D(t) = e^{\Gamma(T)-\Gamma(t)}(J(t) - h(t)),$$

where $J(t)$ is given in Theorem 4.1.

Proof The value of this strategy is

$$
\begin{aligned}
V_\varphi(t) &= \varphi_B(t)B(t,T) + \varphi_D(t)D(t,T) \\
&= h(t \wedge \tau)e^{-r(T-t)} + e^{\Gamma(T)-\Gamma(t)}(J(t) - h(t))\mathbf{1}_{\{t<\tau\}}e^{-r(T-t)}e^{-(\Gamma(T)-\Gamma(t))} \\
&= e^{-r(T-t)}\left(h(t \wedge \tau) + (J(t) - h(t))\mathbf{1}_{\{t<\tau\}}\right) \\
&= e^{-r(T-t)}\left(\mathbf{1}_{\{\tau\le t\}}h(\tau) + \mathbf{1}_{\{t<\tau\}}J(t)\right).
\end{aligned}
$$

On the other hand, Proposition 2.27 gives

$$
\begin{aligned}
\mathbb{E}(h(\tau)|\mathcal{I}_t) &= \mathbf{1}_{\{\tau\le t\}}\mathbb{E}(h(\tau)|\sigma(\tau)) + \mathbf{1}_{\{t<\tau\}}e^{\Gamma(t)}\mathbb{E}(\mathbf{1}_{\{t<\tau\}}h(\tau)) \\
&= \mathbf{1}_{\{\tau\le t\}}h(\tau) + \mathbf{1}_{\{t<\tau\}}J(t).
\end{aligned}
$$

By comparing these two expressions, we obtain

$$
V_\varphi(t) = e^{-r(T-t)}\mathbb{E}(h(\tau)|\mathcal{I}_t)
$$

as claimed.

Furthermore, according to the martingale representation theorem (Theorem 4.1),

$$
\mathbb{E}(h(\tau)|\mathcal{I}_t) = \mathbb{E}(h(\tau)) + \int_0^t (h(s) - J(s))dM(s),
$$

which can be written in terms of the discounted value process $\tilde{V}_\varphi(t) = e^{-rt}V_\varphi(t) = e^{-rT}\mathbb{E}(h(\tau)|\mathcal{I}_t)$ as

$$
\tilde{V}_\varphi(t) = V_\varphi(0) + e^{-rT}\int_0^t (h(s) - J(s))dM(s).
$$

The integration-by-parts formula (Theorem A.9) then gives

$$
\begin{aligned}
V_\varphi(t) - V_\varphi(0) &= e^{rt}\tilde{V}_\varphi(t) - e^{r0}\tilde{V}_\varphi(0) \\
&= \int_0^t \tilde{V}_\varphi(s)d(e^{rs}) + \int_0^t e^{rs}d\tilde{V}_\varphi(s) \\
&= r\int_0^t V_\varphi(s)ds + \int_0^t e^{-r(T-s)}(h(s) - J(s))\,dM(s).
\end{aligned}
$$

Since $V_\varphi(t) = \varphi_B(t)B(t,T) + \varphi_D(t)D(t,T)$, $\varphi_D(t) = e^{\Gamma(T)-\Gamma(t)}(J(t) - h(t))$, and $\hat{D}(t,T) = e^{-r(T-t)}e^{-(\Gamma(T)-\Gamma(t))}$, it follows that

$$
V_\varphi(t) - V_\varphi(0)
$$
$$
= r\int_0^t \varphi_B(s)B(s,T)ds + r\int_0^t \varphi_D(s)D(s,T)ds - \int_0^t \varphi_D(s)\hat{D}(s,T)dM(s).
$$

Finally, since $B(t, T) = e^{-r(T-t)}$ and $D(t, T)$ satisfies (3.8), we obtain the self-financing condition

$$V_\varphi(t) - V_\varphi(0) = \int_0^t \varphi_B(s) dB(s, T) + \int_0^t \varphi_D(s) dD(s, T)$$

by applying the chain rule for Lebesgue–Stieltjes integrals (Theorem A.10). Moreover, the discounted value process $\tilde{V}_\varphi(t) = e^{-rt} V_\varphi(t) = e^{-rT} \mathbb{E}(h(\tau)|\mathcal{I}_t)$ is a martingale, and we can conclude that $\varphi = (\varphi_B, \varphi_D)$ is indeed an admissible self-financing strategy.   □

Exercise 4.8   Provide an alternative argument in the final part of the proof of Theorem 4.11 by computing $\varphi_B(t)$ from (4.4).

Exercise 4.9   Show that the formulae for the components $\varphi_B(t), \varphi_D(t)$ of the replicating strategy in Theorem 4.11 agree with those for a defaultable bond with constant recovery in Example 4.10.

If a recovery payment $\eta(\tau)$ is due at default, then the equivalent payoff at time T will be of the form $h(\tau)$, where

$$h(t) = \mathbf{1}_{\{T < t\}} + \eta(t) e^{r(T-t)} \mathbf{1}_{\{t \leq T\}}.$$

We can apply Theorem 4.11 to find the replicating strategy, and then use it to compute the price $D_\eta(t, T)$ of a defaultable bond with such a recovery payment.

Proposition 4.12
Let η be a left-continuous function such that $\eta(\tau)$ is integrable. The price of a bond paying $\eta(\tau)$ at default is

$$D_\eta(t, T) = \mathbf{1}_{\{\tau \leq t\}} e^{r(t-\tau)} \eta(\tau)$$
$$+ \mathbf{1}_{\{t < \tau\}} \left(e^{-r(T-t)} e^{-(\Gamma(T) - \Gamma(t))} + e^{-r(T-t)} e^{\Gamma(t)} \int_t^T \eta(s) e^{r(T-s)} f(s) ds \right).$$
$$(4.6)$$

Proof   We apply Theorem 4.11 to find the replicating strategy. Let $t \in$

$[0, T]$. First we compute

$$J(t) = e^{\Gamma(t)}\mathbb{E}(h(\tau)\mathbf{1}_{\{t<\tau\}})$$

$$= e^{\Gamma(t)}\mathbb{E}\left(\left(\mathbf{1}_{\{T<\tau\}} + \eta(\tau)e^{r(T-\tau)}\mathbf{1}_{\{\tau\le T\}}\right)\mathbf{1}_{\{t<\tau\}}\right)$$

$$= e^{\Gamma(t)}\mathbb{E}\left(\mathbf{1}_{\{T<\tau\}}\right) + e^{\Gamma(t)}\mathbb{E}\left(\eta(\tau)e^{r(T-\tau)}\mathbf{1}_{\{t<\tau\le T\}}\right)$$

$$= e^{\Gamma(t)-\Gamma(T)} + e^{\Gamma(t)}\int_t^T \eta(s)e^{r(T-s)}f(s)ds.$$

It follows that

$$\varphi_D(t) = e^{\Gamma(T)-\Gamma(t)}(J(t) - h(t))$$

$$= e^{\Gamma(T)-\Gamma(t)}\left(e^{\Gamma(t)-\Gamma(T)} + e^{\Gamma(t)}\int_t^T \eta(s)e^{r(T-s)}f(s)ds\right)$$

$$\quad - e^{\Gamma(T)-\Gamma(t)}\left(\mathbf{1}_{\{T<t\}} + \eta(t)e^{r(T-t)}\mathbf{1}_{\{t\le T\}}\right)$$

$$= 1 + e^{\Gamma(T)}\int_t^T \eta(s)e^{r(T-s)}f(s)ds - \eta(t)e^{\Gamma(T)-\Gamma(t)}e^{r(T-t)}.$$

The default-free position is given by

$$\varphi_B(t) = h(t \wedge \tau)$$

$$= \mathbf{1}_{\{T<t\wedge\tau\}} + \eta(t \wedge \tau)e^{r(T-t\wedge\tau)}\mathbf{1}_{\{t\wedge\tau\le T\}}$$

$$= B(t \wedge \tau, T)^{-1}\eta(t \wedge \tau).$$

If $\tau \le t$, then $D(t, T) = 0$, so

$$D_\eta(t, T) = \varphi_B(t)B(t, T) + \varphi_D(t)D(t, T) = e^{r(t-\tau)}\eta(\tau).$$

If $t < \tau$, then $D(t, T) = e^{-r(T-t)}e^{-(\Gamma(T)-\Gamma(t))}$, so

$$D_\eta(t, T) = \varphi_B(t)B(t, T) + \varphi_D(t)D(t, T)$$

$$= \eta(t) + \left(1 + e^{\Gamma(T)}\int_t^T \eta(s)e^{r(T-s)}f(s)ds\right.$$

$$\left. - \eta(t)e^{\Gamma(T)-\Gamma(t)}e^{r(T-t)}\right)e^{-r(T-t)}e^{-(\Gamma(T)-\Gamma(t))}$$

$$= e^{-r(T-t)}e^{-(\Gamma(T)-\Gamma(t))} + e^{-r(T-t)}e^{\Gamma(t)}\int_t^T \eta(s)e^{r(T-s)}f(s)ds.$$

$\square$

Exercise 4.10 Show that the initial price of a bond paying $\eta(\tau)$ at default is

$$D_\eta(0,T) = e^{-rT}G(T) + \int_0^T \eta(s)e^{-rs}f(s)ds.$$

Exercise 4.11 Derive formula (4.6) for $D_\eta(t,T)$ by computing the conditional expectation of the discounted payoff.

Exercise 4.12 Show that $D_\eta(t,T)$ satisfies the equation

$$dD_\eta(t,T) = \left(rD_\eta(t,T) - \eta(t)\gamma(t)(1 - I(t))\right)dt - \hat{D}_\eta(t,T)dM(t)$$

on $\{t < \tau\}$.

The final example in this section is a defaultable coupon bond. Since it can be regarded as a portfolio containing zero-coupon bonds with maturities shorter than T, we discuss these first.

Example 4.13

We shall price a defaultable unit bond with zero recovery and maturity S, where $0 < S < T$. The payoff due at time S is

$$D(S,S) = \mathbf{1}_{\{S<\tau\}}.$$

To apply the pricing theorem we consider a European claim with payoff

$$h(\tau) = \mathbf{1}_{\{S<\tau\}}e^{r(T-S)}$$

due at time T. Then, according to the general theory, for any $t \in [0,S]$

$$D(t,S) = e^{-r(T-t)}\mathbb{E}(h(\tau)|\mathcal{I}_t)$$
$$= e^{-r(T-t)}\mathbb{E}(\mathbf{1}_{\{S<\tau\}}e^{r(T-S)}|\mathcal{I}_t)$$
$$= e^{-r(S-t)}\mathbb{E}(\mathbf{1}_{\{S<\tau\}}|\mathcal{I}_t).$$

This expectation was computed in Corollary 2.26, hence

$$D(t,S) = \mathbf{1}_{\{t<\tau\}}e^{-r(S-t)}e^{-(\Gamma(S)-\Gamma(t))}.$$

In particular,

$$D(0, S) = e^{-rS} e^{-\Gamma(S)} = e^{-rS} G(S).$$

Exercise 4.13 Find a strategy replicating the bond $D(t, S)$ by means of $B(t, T)$ and $D(t, T)$.

Remark 4.14
As a corollary, we can conclude that $e^{-rt} D(t, S)$ is a martingale under the risk-neutral probability Q. This means that the risk-neutral probability for the bond with maturity T plays the same role for all shorter maturities.

We can immediately obtain a formula for the price of a defaultable coupon bond. Suppose that fixed coupons C_k are paid at times T_k for $k = 1, \ldots, N$, with face value F due at time T_N, where $0 < T_1 < \cdots < T_N$. The initial price of this bond is

$$D_{\text{coupon}}(0) = \sum_{k=1}^{N} C_k D(0, T_k) + F D(0, T_N)$$

$$= \sum_{k=1}^{N} C_k e^{-rT_k} G(T_k) + F e^{-rT_N} G(T_N).$$

Exercise 4.14 Find a strategy replicating this coupon bond by means of $B(t, T)$ and $D(t, T)$ with $T \geq T_N$.

4.3 Credit Default Swap

Suppose that agent A issues a defaultable zero-coupon bond with maturity T and sells it to agent B for $D(0, T)$. The cash flow from A to B will be $\mathbf{1}_{\{T < \tau\}}$ at time T. Now suppose that agent B wishes to purchase protection against the default scenario, and buys it from agent C. At time T, agent C will compensate B in the case of default, so the cash flow from C to B will be $\mathbf{1}_{\{\tau \leq T\}}$ at time T.

How much should B pay C for this service? Since B will receive altogether $\mathbf{1}_{\{T<\tau\}} + \mathbf{1}_{\{\tau \leq T\}} = 1$ unit of cash in each scenario at time T, B has effectively acquired a non-defaultable zero-coupon bond, which costs $B(0, T)$. Hence the premium that B should pay C at time 0 must be equal to the difference $B(0, T) - D(0, T)$.

We can analyse this from the point of view of C. The strategy to replicate the payoff that C needs to deliver to B is to buy the non-defaultable bond and short the defaultable one. This costs $B(0, T) - D(0, T)$, which is exactly how much C is paid by B.

The above is just a hypothetical situation, without much practical relevance. In fact, agent B could have bought a non-defaultable bond in the first place. In reality, the premium is paid by a series of instalments rather than at time 0. This will be analysed below. Here we just note that B does not have to own the defaultable bond issued by A to enter into the contract with C.

The payment $\mathbf{1}_{\{\tau \leq T\}}$ from C to B at time T is called the **default leg**. In practice, the payment from B to C is realised by a series of instalments spread over time, called the **premium leg** or **fixed leg**, and no money changes hands at the initiation of the contract at time 0. We denote these instalments by A_k and assume that they are due at times t_k for $k = 1, \ldots, N$, with $0 < t_1 < \cdots < t_N = T$. These amounts are only paid up to the default time τ. A contract of this kind is known as a **Credit Default Swap** (CDS).

For simplicity, we assume that the A_k are constant and paid at equal time intervals. For $k = 1, \ldots, N$, let $t_k = k\frac{T}{N}$ and let $A_k = \alpha$, called the CDS **spread**. The value of α must be consistent with the condition that no money is paid at time 0. This is similar to interest rate swaps or to forward contracts, where the swap rate or the forward price is determined by the condition that the initial value of the contract should be zero.

To find α we need to equate the time 0 values of both legs of the CDS. The value of the default leg is $\mathbb{E}(e^{-rT} \mathbf{1}_{\{\tau \leq T\}})$, while that of the premium leg is the sum of the values of the instalments paid prior to the default time τ, so we have

$$\mathbb{E}(e^{-rT} \mathbf{1}_{\{\tau \leq T\}}) = \sum_{k=1}^{N} \mathbb{E}(\alpha e^{-rt_k} \mathbf{1}_{\{t_k < \tau\}}).$$

This gives the following formula for the CDS spread:

$$\alpha = \frac{\mathbb{E}(e^{-rT} \mathbf{1}_{\{\tau \leq T\}})}{\sum_{k=1}^{N} \mathbb{E}(e^{-rt_k} \mathbf{1}_{\{t_k < \tau\}})} = \frac{e^{-rT}(1 - G(T))}{\sum_{k=1}^{N} e^{-rt_k} G(t_k)} = \frac{B(0, T) - D(0, T)}{\sum_{k=1}^{N} D(0, t_k)}. \quad (4.7)$$

Remark 4.15

If an investor buys a defaultable bond, paying $D(0, T)$, and then enters into a CDS, this is equivalent to buying a default-free bond in the first place since the difference $B(0, T) - D(0, T)$ is equal to the initial value of the premium payments.

Exercise 4.15 Derive a formula for the spread of a CDS on a bond with constant recovery paid at maturity.

Exercise 4.16 Compute the spread of a CDS on a bond with zero recovery in the case when τ is exponentially distributed with parameter λ under the risk-neutral measure Q. Specifically, assume that $r = 0.05$, $\lambda = 0.02$, $T = 1$, and $N = 12$.

Credit Default Swaps are typically liquid instruments, and the spreads can be used to calibrate the model of credit risk. When the spread α is known for a single CDS contract, formula (4.7) allows us to find the distribution of τ only if it depends on a single parameter, for instance when the hazard rate γ is constant.

Exercise 4.17 Given that $r = 0.06$, $T = 2$, $N = 24$, and $\alpha = 0.1$, compute the hazard rate γ in the case when it is constant. Hence compute the defaultable bond price $D(0, T)$.

If we are given the spreads $\alpha_1, \ldots, \alpha_N$ corresponding to various maturities $t_1, \ldots, t_N$, then the survival probabilities $G(t_k)$ can be computed by solving the system of simultaneous equations

$$\alpha_n = \frac{e^{-rT}(1 - G(t_n))}{\sum_{k=1}^{n} e^{-rt_k} G(t_k)}$$

for $n = 1, \ldots, N$. This in turn allows a piecewise linear hazard function to be constructed.

Exercise 4.18 Construct a piecewise linear hazard function $\Gamma(t)$ given that $T = 1$, $N = 4$, $r = 0.05$, and $\alpha_1 = 0.012$, $\alpha_2 = 0.009$, $\alpha_3 = 0.011$, $\alpha_4 = 0.010$.

The above scheme based on discrete premium payments has a flaw in that it does not take into account the so-called accrued interest. Namely, if default happens at time $\tau \in (t_k, t_{k+1})$, the party who bought the CDS will in practice need to pay a fraction of the instalment A_{k+1} to account for the time elapsed between t_k and τ. This creates some technical difficulties. The accrued interest is ignored by the majority of pricing methods in the literature.

An abstract modification, called a **stylised CDS**, takes care of this and is not difficult to handle. The stylised CDS is composed of:

- the **premium leg** paying continuously at a constant rate α from time 0 up to the time of default τ or maturity T, whichever comes earlier;
- the **default leg** paying 1 unit of cash at maturity T if default occurs before or at time T, and nothing when there is no default before or at time T.

The rate α, called the **stylised CDS spread**, is set so that the time 0 value of the premium leg is the same as that of the default leg.

We can invest the premium leg cash flow into the default-free bond, which gives the time T payoff

$$h_{\mathrm{p}}(\tau) = \int_0^{T \wedge \tau} e^{r(T-u)} \alpha \, du = \frac{\alpha e^{rT} \left(1 - e^{-r(T \wedge \tau)}\right)}{r}$$

for the premium leg. This makes it possible to apply the general pricing scheme to price the premium leg since we have a payment of the form $h(\tau)$ at time T, with

$$h(t) = \frac{\alpha e^{rT} \left(1 - e^{-r(T \wedge t)}\right)}{r}.$$

We also have

$$h_{\mathrm{d}}(\tau) = \mathbf{1}_{\{\tau \le T\}}$$

for the default leg.

Proposition 4.16

The stylised CDS spread is given by

$$\alpha = r \frac{B(0, T) - D(0, T)}{1 - \int_0^T e^{-rs} f(s) \, ds - D(0, T)}.$$

Proof The initial value of the premium leg is

$$e^{-rT}\mathbb{E}(h(\tau)) = \frac{\alpha}{r}\mathbb{E}\left(1 - e^{-r(T \wedge \tau)}\right)$$

$$= \frac{\alpha}{r}\left(1 - \mathbb{E}\left(e^{-r\tau}\mathbf{1}_{\{\tau \le T\}} + e^{-rT}\mathbf{1}_{\{T < \tau\}}\right)\right)$$

$$= \frac{\alpha}{r}\left(1 - \mathbb{E}\left(e^{-r\tau}\mathbf{1}_{\{\tau \le T\}}\right) - e^{-rT}G(T)\right)$$

$$= \frac{\alpha}{r}\left(1 - \int_0^T e^{-rs}f(s)ds - D(0,T)\right).$$

As before, the initial value of the default leg is $B(0,T) - D(0,T)$. It follows that

$$B(0,T) - D(0,T) = \frac{\alpha}{r}\left(1 - \int_0^T e^{-rs}f(s)ds - D(0,T)\right),$$

which gives the result. $\qquad\qquad\qquad\qquad\qquad\qquad\qquad\qquad\qquad\qquad\square$

5

Hazard process model

In Chapter 1 the time of default was endogenous, determined by the behaviour of the process of asset values. In Chapters 2–4 this time was unrelated to the company's operations. In the present chapter the default time τ may be indirectly related to the performance of the company, but we exclude the extreme case when τ is fully determined by the process of asset prices.

5.1   Market model and risk-neutral probability

We expand the default-free part of the market, which in Chapters 2–4 consisted of just a risk-free bond $B(t, T)$ with constant interest rate, by adding a risky asset with values $S(t)$ obeying the Black–Scholes model. The ingredients of the default-free segment of the market will be:

- a risk-free asset represented by a non-defaultable zero-coupon bond

$$B(t, T) = e^{-r(T-t)};$$

- a risky asset with prices $S(t)$ such that

$$dS(t) = \mu S(t)dt + \sigma S(t)dW_P(t)$$

114

for some constants μ and σ, where $W_P(t)$ is a Wiener process on a probability space $(\Omega, \mathcal{F}, P)$, with P being the real-life probability.

The building block of the defaultable segment of the market is as before:

- a defaultable zero-coupon unit bond $D(t, T)$ with zero recovery, whose payoff

$$D(T, T) = \mathbf{1}_{\{T < \tau\}}$$

is determined by a $(0, \infty)$-valued random variable τ defined on $(\Omega, \mathcal{F}, P)$.

Remark 5.1

Let us discuss some specific market components, starting with the defaultable bond issued by an organisation that may go bankrupt, typically a company. In such a case the assumption of zero recovery is unrealistic. In addition, corporate bonds are normally coupon bearing. Such bonds can be replicated by means of zero-coupon zero-recovery bonds. Hence the bond $D(t, T)$, which is not traded, can play the role of an abstract underlying security, whose characteristics can be calibrated by using traded coupon-bearing bonds with positive recovery (or some other traded securities like Credit Default Swaps).

As far as $S(t)$ is concerned, there are various possibilities. A natural choice is a general economic factor representing the state of the economy. This could be realised as a market index. However, companies that may go bankrupt should not be included in the index if the paths of $S(t)$ are to be continuous, since bankruptcy can be related to a jump in the price process. We may consider an index composed of some solid companies which can be assumed never to default. An alternative would be to use companies that are not financed by debt. An all-equity company cannot go bankrupt in the sense of being unable to fulfil their debt obligations. Another possibility is the rate of exchange between the home currency and another currency (or a basket of currencies).

Alternatively, $S(t)$ could represent the fundamental value of the assets of the company issuing the defaultable bond, since the fundamental value of assets would not be affected by a default caused by some exogenous factors. This, however, would not be fully consistent with the assumption that $S(t)$ represents traded assets.

In a similar manner as in Section 2.3, we define the default indicator process $I(t) = \mathbf{1}_{\{\tau \leq t\}}$ and the filtration $(\mathcal{I}_t)_{t \geq 0}$ generated by the family of random variables $I(t)$, $t \geq 0$. Let $(\mathcal{F}_t)_{t \geq 0}$ be the filtration generated by the Wiener process $W_P(t)$, $t \geq 0$ (or, equivalently, by $S(t)$, $t \geq 0$). Moreover, let

$$\mathcal{G}_t = \sigma(\mathcal{F}_t \cup \mathcal{I}_t)$$

be the **enlarged filtration** covering the information contained in both the default-free and defaultable segments of the market.

The processes $S(t)$ and $I(t)$ are adapted to the filtration $(\mathcal{G}_t)_{t\geq0}$, but $S(t)$ does not have to be adapted to $(\mathcal{I}_t)_{t\geq0}$, nor does $I(t)$ have to be adapted to $(\mathcal{F}_t)_{t\geq0}$, that is, in other words, τ does not have to be an $(\mathcal{F}_t)_{t\geq0}$-stopping time. But, if τ were, for example, defined as in the barrier model of Section 1.4, with $S(t)$ being the process of asset values, then we would have $\mathcal{I}_t \subset \mathcal{F}_t$ and hence $\mathcal{G}_t = \mathcal{F}_t$ for all $t \geq 0$.

Exercise 5.1   Show that $(\mathcal{G}_t)_{t\geq0}$ is the smallest filtration such that τ is a $(\mathcal{G}_t)_{t\geq0}$-stopping time and $\mathcal{F}_t \subset \mathcal{G}_t$ for each $t \geq 0$.

Next we introduce the following fundamental assumption. It is similar to Assumption 3.1, but has broader scope and covers both segments of the market.

Assumption 5.2

There exists a probability measure Q equivalent to P, such that the discounted stock and defaultable bond prices $\tilde{S}(t) = e^{-rt}S(t)$ and $\tilde{D}(t,T) = e^{-rt}D(t,T)$ are $(\mathcal{G}_t)_{t\geq0}$-martingales under Q for each maturity date $T > 0$. Such a Q is called a **risk-neutral probability** or **martingale measure**.

In general, the probability Q satisfying the conditions of Assumption 5.2 may not be unique. We are not going to be worried about this as the risk-neutral probability is chosen by the market, which dictates the prices of financial securities. In particular, the bonds $D(t,T)$ are priced by the market (indirectly, given that only certain derivative securities like defaultable coupon-bearing bonds or Credit Default Swaps are traded). From these prices we can extract the characteristics of the distribution of τ under Q.

In what follows the expectation under Q will be denoted simply by $\mathbb{E}$ without referencing the measure explicitly.

5.2 Martingales with respect to the enlarged filtration

Our next task is to discuss the relationship between martingales with respect to the filtrations $(\mathcal{F}_t)_{t\geq0}$ and $(\mathcal{G}_t)_{t\geq0}$. We begin with a simple observation.

Proposition 5.3
Any $(\mathcal{G}_t)_{t\geq0}$-martingale adapted to the filtration $(\mathcal{F}_t)_{t\geq0}$ is also an $(\mathcal{F}_t)_{t\geq0}$-martingale.

Proof   Let $0 \leq s \leq t$. Suppose that X is a $(\mathcal{G}_t)_{t\geq0}$-martingale adapted to the filtration $(\mathcal{F}_t)_{t\geq0}$. Then

$$\mathbb{E}(X(t)|\mathcal{F}_s) = \mathbb{E}(\mathbb{E}(X(t)|\mathcal{G}_s)|\mathcal{F}_s) \qquad \text{by the tower property}$$
$$= \mathbb{E}(X(s)|\mathcal{F}_s) \qquad\qquad \text{by the definition of martingale}$$
$$= X(s) \qquad\qquad\qquad \text{by adaptedness.}$$

$\square$

Recall (see [BSM]) that for the Black–Scholes market model consisting of a non-defaultable bond $B(t, T) = e^{-r(T-t)}$ (or, equivalently, the money market account $A(t) = e^{rt}$; see Remark 2.30) and a risky asset $S(t)$, there exists a probability measure Q_{BS} such that the process of discounted prices $\tilde{S}(t) = e^{-rt}S(t)$ is an $(\mathcal{F}_t)_{t\geq0}$-martingale under Q_{BS}. Moreover, such a probability Q_{BS} is unique on $\mathcal{F}_t$ for each $t \geq 0$ since the Black–Scholes model is complete.

Corollary 5.4
The probability Q coincides with Q_{BS} on $\mathcal{F}_t$ for any $t > 0$.

Proof   By Assumption 5.2, the discounted stock price $\tilde{S}(t) = e^{-rt}S(t)$ is a $(\mathcal{G}_t)_{t\geq0}$-martingale under Q. Moreover, $\tilde{S}(t)$ is adapted to the filtration $(\mathcal{F}_t)_{t\geq0}$, hence it is an $(\mathcal{F}_t)_{t\geq0}$-martingale by Proposition 5.3. The claim follows from the uniqueness of the probability Q_{BS} in the Black–Scholes model.

$\square$

The converse of Proposition 5.3 is a non-trivial property. It is restricted to $(\mathcal{F}_t)_{t\geq0}$-martingales that are square integrable in the following sense.

Definition 5.5
We say that $X(t), t \geq 0$, is a **square-integrable** $(\mathcal{F}_t)_{t\geq0}$-martingale if there is an $\mathcal{F}_\infty$-measurable random variable Y such that $\mathbb{E}(Y^2) < \infty$ and $X(t) = \mathbb{E}(Y|\mathcal{F}_t)$ for each $t \geq 0$, where

$$\mathcal{F}_\infty = \sigma(\textstyle\bigcup_{t\geq0} \mathcal{F}_t)$$

denotes the smallest σ-field that contains $\mathcal{F}_t$ for each $t \geq 0$.

The property in question, known as Hypothesis (H), can be formulated as follows.

Hypothesis (H)

Any square-integrable $(\mathcal{F}_t)_{t \geq 0}$-martingale is a $(\mathcal{G}_t)_{t \geq 0}$-martingale.

In the market model introduced in the present chapter this property can be proved. The proof involves some knowledge of integration with respect to martingales, outlined in Appendix A.3. Readers who do not feel comfortable with this material can skip it and accept Hypothesis (H) as an assumption.

Theorem 5.6

Hypothesis (H) *holds in the market model consisting of a non-defaultable bond* $B(t, T) = e^{-r(T-t)}$, *a Black–Scholes risky asset* $S(t)$, *and a defaultable bond* $D(t, T)$.

Proof Suppose that X is a square-integrable $(\mathcal{F}_t)_{t \geq 0}$-martingale under Q. By Corollary 5.4, it is also a square-integrable $(\mathcal{F}_t)_{t \geq 0}$-martingale under Q_{BS}. Take any $T > 0$. Then $X(T)$ is an $\mathcal{F}_T$-measurable random variable, square integrable under Q_{BS}. We assume, additionally, that $X(T) \geq 0$. This assumption will be relaxed at the end of the proof. It follows that $X(T)$ can be replicated in the default-free segment (i.e. in the Black–Scholes model) by an admissible self-financing strategy (φ_B, φ_S); see [BSM]. We denote the values of this strategy by

$$V_\varphi(t) = \varphi_B(t)B(t, T) + \varphi_S(t)S(t).$$

The discounted value process $\tilde{V}_\varphi(t) = e^{-rt}V_\varphi(t)$ is an $(\mathcal{F}_t)_{t \geq 0}$-martingale under Q_{BS}, hence under Q.

By replication, $V_\varphi(T) = X(T)$, so

$$\tilde{V}_\varphi(t) = \mathbb{E}(\tilde{V}_\varphi(T)|\mathcal{F}_t) = e^{-rT}\mathbb{E}(X(T)|\mathcal{F}_t) \geq 0$$

for all $t \in [0, T]$. Moreover, because $\tilde{V}_\varphi$ and X are $(\mathcal{F}_t)_{t \geq 0}$-martingales under Q and

$$\tilde{V}_\varphi(T) = e^{-rT}X(T),$$

it follows that

$$\tilde{V}_\varphi(t) = \mathbb{E}(\tilde{V}_\varphi(T)|\mathcal{F}_t) = e^{-rT}\mathbb{E}(X(T)|\mathcal{F}_t) = e^{-rT}X(t)$$

for each $t \in [0, T]$.

The strategy (φ_B, φ_S) is self-financing, so the discounted values satisfy (see [BSM])

$$\tilde{V}_\varphi(t) = \tilde{V}_\varphi(0) + \int_0^t \varphi_S(u)\sigma\tilde{S}(u)dW_Q(u), \tag{5.1}$$

where

$$W_Q(t) = \frac{\mu - r}{\sigma}t + W_P(t) \tag{5.2}$$

is a Wiener process under Q_{BS}, hence under Q, with respect to the filtration $(\mathcal{F}_t)_{t \geq 0}$. Since

$$\tilde{S}(t) = \tilde{S}(0) + \int_0^t \sigma \tilde{S}(u)dW_Q(u)$$

and $\tilde{S}$ is an $(\mathcal{F}_t)_{t \geq 0}$-martingale under Q with almost surely continuous paths, it follows by Theorem A.34 that the integral in (5.1) can be expressed as a stochastic integral with respect to $\tilde{S}$,

$$\tilde{V}_\varphi(t) = \tilde{V}_\varphi(0) + \int_0^t \varphi_S(u)d\tilde{S}(u). \tag{5.3}$$

This stochastic integral is defined under the filtration $(\mathcal{F}_t)_{t \geq 0}$, and the discounted value process $\tilde{V}_\varphi$ is an $(\mathcal{F}_t)_{t \geq 0}$-martingale under Q.

We claim that $\tilde{V}_\varphi$ is also a martingale under Q with respect to the enlarged filtration $(\mathcal{G}_t)_{t \geq 0}$. According to Assumption 5.2, $\tilde{S}$ is a martingale under Q with respect to $(\mathcal{G}_t)_{t \geq 0}$. Hence, by Remark A.33, the stochastic integral on the right-hand side of (5.3) remains unchanged when the filtration $(\mathcal{F}_t)_{t \geq 0}$ is replaced by $(\mathcal{G}_t)_{t \geq 0}$. It follows that this stochastic integral, hence $\tilde{V}_\varphi$, is a local $(\mathcal{G}_t)_{t \geq 0}$-martingale under Q. Because $\tilde{V}_\varphi$ is non-negative and satisfies

$$\mathbb{E}(\tilde{V}_\varphi(t)) = \mathbb{E}(\mathbb{E}(\tilde{V}_\varphi(T)|\mathcal{F}_t)) = \mathbb{E}(\tilde{V}_\varphi(T))$$

for each $t \in [0, T]$, it follows by Proposition A.22 that $\tilde{V}_\varphi$ is indeed a $(\mathcal{G}_t)_{t \geq 0}$-martingale under Q.

We can now relax the assumption that $X(T) \geq 0$ by replicating the positive and negative parts $X(T)^+$ and $X(T)^-$ by two strategies φ^+ and φ^-, respectively. Because $\tilde{V}_{\varphi^+}$ and $\tilde{V}_{\varphi^-}$ are $(\mathcal{F}_t)_{t \geq 0}$-martingales under Q, so is the difference $\tilde{V}_{\varphi^+} - \tilde{V}_{\varphi^-}$. Moreover, since

$$\tilde{V}_{\varphi^+}(T) - \tilde{V}_{\varphi^-}(T) = e^{-rT}X(T)^+ - e^{-rT}X(T)^- = e^{-rT}X(T)$$

and X is also an $(\mathcal{F}_t)_{t \geq 0}$-martingale under Q, it follows that

$$\begin{aligned}
\tilde{V}_{\varphi^+}(t) - \tilde{V}_{\varphi^-}(t) &= \mathbb{E}(\tilde{V}_{\varphi^+}(T) - \tilde{V}_{\varphi^-}(T)|\mathcal{F}_t) \\
&= e^{-rT}\mathbb{E}(X(T)|\mathcal{F}_t) \\
&= e^{-rT}X(t)
\end{aligned}$$

for each $t \in [0, T]$. Finally, it has been proved above that $\tilde{V}_{\varphi^+}$ and $\tilde{V}_{\varphi^-}$ are

$(\mathcal{G}_t)_{t\geq 0}$-martingales under Q, so we can conclude that the process

$$X(t) = e^{rT}(\tilde{V}_{\varphi^+}(t) - \tilde{V}_{\varphi^-}(t))$$

is a $(\mathcal{G}_t)_{t\geq 0}$-martingale under Q.   □

Remark 5.7
The proof that Hypothesis (H) holds would be much simpler if we assumed, as some authors do, that the discounted prices of any tradeable contingent claim is a martingale. Such an assumption would be extremely strong, making most of the theory redundant. For instance, in the Black–Scholes model it would give the Black–Scholes formula trivially in one line. We emphasise that here we only assume the martingale property of $\tilde{S}(t)$ and $\tilde{D}(t,T)$.

Remark 5.8
To show the martingale property of $\tilde{V}_\varphi$ in the proof of Theorem 5.6 we could not avoid referring to the theory of stochastic integration with respect to martingales by arguing that

$$\int_0^t \varphi_S(u)d\tilde{S}(u) = \int_0^t \varphi_S(u)\sigma\tilde{S}(u)dW_Q(u)$$

is a local $(\mathcal{G}_t)_{t\geq 0}$-martingale because we did not know if W_Q is a Wiener process with respect to the filtration $(\mathcal{G}_t)_{t\geq 0}$ in the sense of Definition A.19 in Appendix A.3. As we shall see, the latter is in fact as a consequence of Theorem 5.6.

Corollary 5.9
The process W_Q is a Wiener process with respect to $(\mathcal{G}_t)_{t\geq 0}$ under Q.

Proof   Since W_Q is a Wiener process with respect to the filtration $(\mathcal{F}_t)_{t\geq 0}$ under Q, the processes $W_Q(t)$ and $W_Q(t)^2 - t$ for $t \geq 0$ are $(\mathcal{F}_t)_{t\geq 0}$-martingales under Q. Take any $T > 0$. Because $W_Q(T)$ and $W_Q(T)^2 - T$ are square-integrable random variables under Q, it follows that $W_Q(t)$ and $W_Q(t)^2 - t$ for $t \in [0,T]$ are square-integrable $(\mathcal{F}_t)_{t\geq 0}$-martingales under Q. By Theorem 5.6, they are therefore $(\mathcal{G}_t)_{t\geq 0}$-martingales under Q. This is so for any $T > 0$, hence $W_Q(t)$ and $W_Q(t)^2 - t$ defined for all $t \geq 0$ are $(\mathcal{G}_t)_{t\geq 0}$-martingales under Q. According to Lévy's characterisation of the Wiener process, Theorem A.23 in Appendix A.3, W_Q is therefore a Wiener process with respect to $(\mathcal{G}_t)_{t\geq 0}$ under Q.   □

5.3 Hazard process

The relationship between the default-free and defaultable segments of the market can be captured by the following process.

Definition 5.10
The **cumulative distribution process** for the default time τ is given by

$$F(t) = Q(\tau \leq t | \mathcal{F}_t)$$

for each $t \geq 0$.

We are using similar notation as for the cumulative distribution function in Chapter 2, but $F(t)$ is now a stochastic process. The right-hand side can also be written as a conditional expectation,

$$F(t) = Q(\tau \leq t | \mathcal{F}_t) = \mathbb{E}(\mathbf{1}_{\{\tau \leq t\}} | \mathcal{F}_t).$$

Remark 5.11
Since $\tau > 0$, we have $F(0) = 0$. Moreover, since $\tau < \infty$, $\lim_{t \to \infty} F(t) = 1$ almost surely because $\lim_{t \to \infty} \mathbf{1}_{\{\tau \leq t\}} = 1$.

We want to have a non-trivial relationship between S and D, but first consider two extreme cases.

Example 5.12
Suppose that τ is a stopping time with respect to the filtration $(\mathcal{F}_t)_{t \geq 0}$. This property can be expressed as $\mathcal{I}_t \subset \mathcal{F}_t$. It will take place if τ is defined by means of $S(t)$, for instance as the first time when $S(t)$ crosses a given deterministic barrier. Then, since $\mathbf{1}_{\{\tau \leq t\}}$ is $\mathcal{F}_t$-measurable, $F(t) = \mathbf{1}_{\{\tau \leq t\}} = I(t)$. We find ourselves within the framework of the barrier model (Section 1.4), where the stock prices $S(t)$ determine τ, with undesirable features like zero short-time spreads or predictability of τ.

Example 5.13
Suppose that τ is independent of all $\mathcal{F}_t$ (under Q). Then $\mathbb{E}(\mathbf{1}_{\{\tau \leq t\}} | \mathcal{F}_t) = \mathbb{E}(\mathbf{1}_{\{\tau \leq t\}}) = Q(\tau \leq t)$, hence $F(t)$ is the ordinary cumulative distribution function of τ, and the theory developed in Chapters 2–4 provides all that is necessary. If S is independent of τ, it is irrelevant to the behaviour of

the defaultable bond. This case is not particularly interesting, but it is not excluded in what follows.

We formulate an assumption, which we also imposed before in the simplified model of Chapter 2; see Assumption 2.3. It guarantees that the key object (the hazard process, see below) can be defined.

Assumption 5.14

$F(t) < 1$ for all $t > 0$.

If τ is an $(\mathcal{F}_t)_{t\geq 0}$-stopping time, then this condition is not satisfied since then $F(t) = \mathbf{1}_{\{\tau \leq t\}}$ is equal to 1 with positive probability for some $t > 0$. This, in particular, means that a model satisfying Assumption 5.14 has to be incompatible with the barrier model of Section 1.4.

A consequence of Assumption 5.14 is that τ cannot be an $(\mathcal{F}_t)_{t\geq 0}$-stopping time. In other words, the filtration $(\mathcal{G}_t)_{t\geq 0}$ contains non-trivial additional information compared to $(\mathcal{F}_t)_{t\geq 0}$. Observe that since τ is an $(\mathcal{I}_t)_{t\geq 0}$-stopping time, it is a $(\mathcal{G}_t)_{t\geq 0}$-stopping time (because $\mathcal{I}_t \subset \mathcal{G}_t$).

The process $F(t)$ has the following property, a natural extension of the fact that the cumulative distribution function is non-decreasing.

Exercise 5.2 Show that $F(t)$ is a submartingale with respect to the filtration $(\mathcal{F}_t)_{t\geq 0}$.

The next property shows that the conditional probability of default at a given time $s \geq 0$ is not affected by information about events at a later time $t \geq s$, which agrees with intuition.

Lemma 5.15

For any $0 \leq s \leq t$ *we have*

$$F(s) = \mathbb{E}(\mathbf{1}_{\{\tau \leq s\}}|\mathcal{F}_t).$$

Proof Take any $A \in \mathcal{F}_t$. First we show that for any $s \leq t$

$$\mathbb{E}(\mathbf{1}_A|\mathcal{F}_s) = \mathbb{E}(\mathbf{1}_A|\mathcal{G}_s). \qquad (5.4)$$

The process $\mathbb{E}(\mathbf{1}_A|\mathcal{F}_t)$ is a square-integrable $(\mathcal{F}_t)_{t\geq 0}$-martingale, so it is a $(\mathcal{G}_t)_{t\geq 0}$-martingale by Theorem 5.6,

$$\mathbb{E}(\mathbb{E}(\mathbf{1}_A|\mathcal{F}_t)|\mathcal{G}_s) = \mathbb{E}(\mathbf{1}_A|\mathcal{F}_s).$$

The random variable $\mathbf{1}_A$ is $\mathcal{F}_t$-measurable, so $\mathbb{E}(\mathbf{1}_A|\mathcal{F}_t) = \mathbf{1}_A$. Hence

$$\mathbb{E}(\mathbb{E}(\mathbf{1}_A|\mathcal{F}_t)|\mathcal{G}_s) = \mathbb{E}(\mathbf{1}_A|\mathcal{G}_s),$$

which proves (5.4).

To demonstrate the lemma we need to show that for any $A \in \mathcal{F}_t$

$$\mathbb{E}(\mathbf{1}_A\mathbf{1}_{\{\tau \le s\}}) = \mathbb{E}(\mathbf{1}_A F(s)),$$

which, after inserting the definition of $F(s)$, becomes

$$\mathbb{E}(\mathbf{1}_A\mathbf{1}_{\{\tau \le s\}}) = \mathbb{E}(\mathbf{1}_A\mathbb{E}(\mathbf{1}_{\{\tau \le s\}}|\mathcal{F}_s)).$$

We transform the expectation on the left:

$$
\begin{aligned}
\mathbb{E}(\mathbf{1}_A\mathbf{1}_{\{\tau \le s\}}) &= \mathbb{E}(\mathbb{E}(\mathbf{1}_A\mathbf{1}_{\{\tau \le s\}}|\mathcal{G}_s)) && \text{by the tower property} \\
&= \mathbb{E}(\mathbf{1}_{\{\tau \le s\}}\mathbb{E}(\mathbf{1}_A|\mathcal{G}_s)) && \text{since } \mathcal{I}_s \subset \mathcal{G}_s \\
&= \mathbb{E}(\mathbf{1}_{\{\tau \le s\}}\mathbb{E}(\mathbf{1}_A|\mathcal{F}_s)) && \text{by (5.4)} \\
&= \mathbb{E}(\mathbb{E}(\mathbf{1}_{\{\tau \le s\}}\mathbb{E}(\mathbf{1}_A|\mathcal{F}_s)|\mathcal{F}_s)) && \text{by the tower property} \\
&= \mathbb{E}(\mathbb{E}(\mathbf{1}_A|\mathcal{F}_s)\mathbb{E}(\mathbf{1}_{\{\tau \le s\}}|\mathcal{F}_s)) && \text{since } \mathbb{E}(\mathbf{1}_A|\mathcal{F}_s) \text{ is } \mathcal{F}_s\text{-measurable} \\
&= \mathbb{E}(\mathbb{E}(\mathbf{1}_A\mathbb{E}(\mathbf{1}_{\{\tau \le s\}}|\mathcal{F}_s)|\mathcal{F}_s)) && \text{since } \mathbb{E}(\mathbf{1}_{\{\tau \le s\}}|\mathcal{F}_s) \text{ is } \mathcal{F}_s\text{-measurable} \\
&= \mathbb{E}(\mathbf{1}_A\mathbb{E}(\mathbf{1}_{\{\tau \le s\}}|\mathcal{F}_s)) && \text{by the tower property.}
\end{aligned}
$$

This completes the argument. □

Corollary 5.16

The cumulative distribution process of τ can be expressed as

$$F(t) = \mathbb{E}(\mathbf{1}_{\{\tau \le t\}}|\mathcal{F}_\infty).$$

Proof We need to show that

$$\mathbb{E}(\mathbf{1}_{\{\tau \le t\}}|\mathcal{F}_\infty) = \mathbb{E}(\mathbf{1}_{\{\tau \le t\}}|\mathcal{F}_t),$$

that is, for any $A \in \mathcal{F}_\infty$,

$$\mathbb{E}(\mathbf{1}_A\mathbf{1}_{\{\tau \le t\}}) = \mathbb{E}(\mathbf{1}_A\mathbb{E}(\mathbf{1}_{\{\tau \le t\}}|\mathcal{F}_t)). \tag{5.5}$$

To this end take $A \in \bigcup_{u \ge 0} \mathcal{F}_u$, a family of sets which generates $\mathcal{F}_\infty$. Hence $A \in \mathcal{F}_u$ for some $u \ge 0$. If $u \le t$, then $A \in \mathcal{F}_t$, and the above equality follows from the tower property since we have $\mathbb{E}(\mathbb{E}(\mathbf{1}_A\mathbf{1}_{\{\tau \le t\}}|\mathcal{F}_t))$ on the right. If $u > t$, then $\mathbb{E}(\mathbf{1}_{\{\tau \le t\}}|\mathcal{F}_t) = \mathbb{E}(\mathbf{1}_{\{\tau \le t\}}|\mathcal{F}_u)$ by Lemma 5.15, so on the right-hand side we have $\mathbb{E}(\mathbb{E}(\mathbf{1}_A\mathbf{1}_{\{\tau \le t\}}|\mathcal{F}_u))$, and the tower property can be applied once again.

To extend the result to the whole σ-field $\mathcal{F}_\infty$, it is routine to note that either side of (5.5) is a countably additive function of A. Hence the family of sets A such that (5.5) holds is a σ-field, and so it contains $\mathcal{F}_\infty$. □

Exercise 5.3   Use Lemma 5.15 to show that the process $F(t)$ is non-decreasing.

It is convenient to define the **survival process** as

$$G(t) = 1 - F(t).$$

By the linearity of conditional expectation, $G(t) = Q(t < \tau | \mathcal{F}_t)$, so this process describes the conditional probability of survival.

Our next step is to define a natural counterpart of the hazard function.

Definition 5.17
The **hazard process** is defined as

$$\Gamma(t) = -\ln G(t) = -\ln(1 - F(t)).$$

This is well defined because $G(t) > 0$ by Assumption 5.14. Since $Q(\tau \le 0) = 0$, it follows that $\Gamma(0) = 0$.

Exercise 5.4   Show that $\Gamma(t)$ is a submartingale with respect to the filtration $(\mathcal{F}_t)_{t \ge 0}$.

Similarly as in the earlier chapters, we impose an additional regularity assumption to simplify the theory.

Assumption 5.18
The paths of $F(t)$ are strictly increasing and absolutely continuous, that is, $F(t) = \int_0^t f(s)ds$ for some process $f(t) > 0$.

This assumption allows us to perform pathwise integration with respect to $F(t)$. Note that $f(t)$ is a process, not a density (though each path can be regarded as one).

The regularity of the paths of Γ follows from the regularity of the paths of F. The paths of Γ are continuous and increasing. Since $\lim_{t \to \infty} F(t) = 1$ almost surely, it follows that $\lim_{t \to \infty} \Gamma(t) = \infty$ almost surely, though these asymptotic properties are not important as we are mainly interested in a finite time interval $[0, T]$.

Definition 5.19
If $\Gamma(t)$ has absolutely continuous paths, the **hazard rate process** $\gamma(t)$ is

defined by

$$\Gamma(t) = \int_0^t \gamma(u)du.$$

Proposition 5.20

The hazard rate process exists and it is given by

$$\gamma(t) = \frac{f(t)}{1 - F(t)}.$$

Moreover, $f(t)$ can be expressed by means of $\gamma(t)$ as

$$f(t) = \gamma(t)G(t) = \gamma(t)e^{-\int_0^t \gamma(u)du}.$$

Proof This is an immediate consequence of Proposition 2.13 applied pathwise. $\square$

5.4 Canonical construction of default time

We are going to invert the above constructions and build τ by starting with a given process Γ.

Assumption 5.21

Let $\Gamma(t), t \geq 0$, be an $(\mathcal{F}_t)_{t\geq0}$-adapted process with continuous increasing paths such that $\Gamma(0) = 0$ and $\lim_{t\to\infty} \Gamma(t) = \infty$.

The process $e^{-\Gamma(t)}$ takes values in the unit interval $[0, 1]$, starts with $e^{-\Gamma(0)} = 1$, and converges monotonically to zero as $t \to \infty$. The set $\{t \geq 0 : e^{-\Gamma(t)} \leq x\}$ is non-empty for any $x \in (0, 1]$, so the random time $\inf\{t \geq 0 : e^{-\Gamma(t)} \leq x\}$ is finite. Such a random variable is an $\mathcal{F}_t$-stopping time, a situation we want to avoid. Hence we replace x by a random variable.

Let X be a random variable independent of $\mathcal{F}_\infty = \sigma(\bigcup_{t\geq0} \mathcal{F}_t)$ and uniformly distributed in $[0, 1]$. We define

$$\tau = \inf\{t \geq 0 : e^{-\Gamma(t)} \leq X\}.$$

This makes it possible, in particular, to simulate the default time numerically. It is sufficient to generate a path of the process Γ, and independently draw a random number from the unit interval.

Example 5.22

We can produce an example based on the process $S(t), t \geq 0$, of Black–Scholes stock prices. Consider the logarithmic return

$$\ln \frac{S(t)}{S(0)} = \left(r - \frac{1}{2}\sigma^2\right)t + \sigma W_Q(t),$$

where $W_Q(t)$ is a Wiener process under the risk-neutral probability Q, and let

$$\Gamma(t) = \max_{u \in [0,t]} \left(-\ln \frac{S(u)}{S(0)}\right).$$

It follows that $e^{-\Gamma(t)} \leq X$ when $\min_{u \in [0,t]} S(u) \leq XS(0)$. Default is triggered when the stock price falls below the random fraction X of the initial value.

Exercise 5.5   What is the probability distribution of the default time

$$\tau = \inf\{t \geq 0 : e^{-\Gamma(t)} \leq X\}$$

in Example 5.22?
   Hint: Use Lemma A.16 in Appendix A.2.

A more direct alternative is to consider a random variable Y independent of $\mathcal{F}_\infty$ and having the unit exponential distribution, that is, with the cumulative distribution function

$$F_Y(t) = \begin{cases} 1 - e^{-t} & \text{for } t > 0, \\ 0 & \text{otherwise,} \end{cases}$$

and define

$$\tau = \inf\{t \geq 0 : \Gamma(t) \geq Y\}. \tag{5.6}$$

Exercise 5.6   Show that e^{-Y} is uniformly distributed in $[0, 1]$ if and only if Y has the unit exponential distribution.

> **Exercise 5.7** Show that the two definitions (5.4) and (5.6) of τ produce random variables with the same distribution.

The constructions are clearly equivalent when $X = e^{-Y}$.

Proposition 5.23
The hazard process of τ is equal to the prescribed process $\Gamma(t)$, that is,

$$Q(\tau \le t) = 1 - e^{-\Gamma(t)}.$$

Proof This can be verified as follows:

$$
\begin{aligned}
Q(\tau \le t) &= \mathbb{E}(\mathbf{1}_{\{\tau \le t\}}|\mathcal{F}_t) & \text{by the definition of } F(t) \\
&= \mathbb{E}(\mathbb{E}(\mathbf{1}_{\{\tau \le t\}}|\mathcal{F}_\infty)|\mathcal{F}_t) & \text{by the tower property} \\
&= \mathbb{E}(\mathbb{E}(\mathbf{1}_{\{Y \le \Gamma(t)\}}|\mathcal{F}_\infty)|\mathcal{F}_t) & \text{by the definition of } \tau \\
&= \mathbb{E}(F_Y(\Gamma(t))|\mathcal{F}_t) & \text{since } Y, \mathcal{F}_\infty \text{ are independent} \\
&= 1 - e^{-\Gamma(t)} & \text{since } \Gamma \text{ is } (\mathcal{F}_t)_{t \ge 0}\text{-adapted.}
\end{aligned}
$$

$\square$

Remark 5.24
The construction of τ can start with a hazard rate process prescribed in advance. Let $\gamma(t) > 0$ be a process adapted to the filtration $(\mathcal{F}_t)_{t \ge 0}$, and take $\Gamma(t) = \int_0^t \gamma(u)du$. For Γ to have all the properties in Assumption 5.21 we require that $\int_0^\infty \gamma(t)dt = \infty$.

The next proposition shows that, with no loss of generality, we can assume that any default time such that the corresponding hazard process $\Gamma(t)$ has continuous increasing paths comes from the canonical construction with a suitably chosen random variable Y. We precede it by a general lemma.

Lemma 5.25
For any random variable X and for any σ-field $\mathcal{G}$, if $\mathbb{E}(\mathbf{1}_{\{X \le t\}}|\mathcal{G})$ is deterministic for each $t \in \mathbb{R}$, then X and $\mathcal{G}$ are independent.

Proof Let $B \in \mathcal{G}$, and take $A = \{X \le t\}$ for some $t \in \mathbb{R}$. Then

$$
\begin{aligned}
\mathbb{E}(\mathbf{1}_{\{X \le t\}}\mathbf{1}_B) &= \mathbb{E}(\mathbb{E}(\mathbf{1}_{\{X \le t\}}|\mathcal{G})\mathbf{1}_B) & \text{by the definition of } \mathbb{E}(\mathbf{1}_{\{X \le t\}}|\mathcal{G}) \\
&= \mathbb{E}(\mathbf{1}_{\{X \le t\}}|\mathcal{G})Q(B) & \text{since } \mathbb{E}(\mathbf{1}_{\{X \le t\}}|\mathcal{G}) \text{ is deterministic.}
\end{aligned}
$$

After taking the expectation on both sides, we get

$$Q(A \cap B) = Q(A)Q(B) \tag{5.7}$$

for all sets A of the form $A = \{X \leq t\}$, where $t \in \mathbb{R}$, which generate $\sigma(X)$. The family of sets A that satisfy (5.7) is clearly a σ-field, hence it contains $\sigma(X)$. This means that (5.7) holds for every $A \in \sigma(X)$ and $B \in \mathcal{G}$, so X and $\mathcal{G}$ are independent.   □

Proposition 5.26

If τ is such that the corresponding hazard process $\Gamma(t)$ has continuous increasing paths, then there exists a random variable Y with the unit exponential distribution and independent of $\mathcal{F}_\infty$ such that

$$\tau = \inf\{t \geq 0 : \Gamma(t) \geq Y\}.$$

Proof Since the process $\Gamma(t)$ has increasing paths,

$$\tau = \inf\{t \geq 0 : \Gamma(t) \geq \Gamma(\tau)\}.$$

This suggests taking $Y = \Gamma(\tau)$. We need to show that $\Gamma(\tau)$ has the unit exponential distribution and is independent of $\mathcal{F}_\infty$.

By Corollary 5.16, for each $t \geq 0$

$$\mathbb{E}(\mathbf{1}_{\{\tau \leq t\}} | \mathcal{F}_\infty) = F(t) = 1 - e^{-\Gamma(t)}.$$

We denote by Γ^{-1} the process whose paths are the inverse functions to those of Γ. It is well defined because the paths of Γ are increasing. For any $t \geq 0$, observe that $\Gamma^{-1}(t) \geq 0$ and $\{\Gamma^{-1}(t) \leq u\} = \{t \leq \Gamma(u)\}$ for each $u \geq 0$. It follows that $\Gamma^{-1}(t)$ is $\mathcal{F}_\infty$-measurable since $\Gamma(u)$ is $\mathcal{F}_u$-measurable, hence $\mathcal{F}_\infty$-measurable, for each $u \geq 0$.

Next, we show that, for every $\mathcal{F}_\infty$-measurable simple random variable

$$X = \sum_{i=1}^{I} x_i \mathbf{1}_{A_i},$$

where $x_i \geq 0$ and $A_i \in \mathcal{F}_\infty$ for each $i = 1, \ldots, I$, we have

$$\mathbb{E}(\mathbf{1}_{\{\tau \leq X\}} | \mathcal{F}_\infty) = F(X).$$

We can assume without loss of generality that the events A_i are pairwise disjoint. Then, since $F(0) = 0$ and $\tau > 0$, we get $\mathbf{1}_{\{\tau \leq X\}} = \sum_{i=1}^{I} \mathbf{1}_{\{\tau \leq x_i\}} \mathbf{1}_{A_i}$ and $F(X) = \sum_{i=1}^{I} F(x_i) \mathbf{1}_{A_i}$. It follows that

$$\mathbb{E}(\mathbf{1}_{\{\tau \leq X\}} | \mathcal{F}_\infty) = \sum_{i=1}^{I} \mathbb{E}(\mathbf{1}_{\{\tau \leq x_i\}} \mathbf{1}_{A_i} | \mathcal{F}_\infty) = \sum_{i=1}^{I} \mathbb{E}(\mathbf{1}_{\{\tau \leq x_i\}} | \mathcal{F}_\infty) \mathbf{1}_{A_i}$$

$$= \sum_{i=1}^{I} F(x_i) \mathbf{1}_{A_i} = F(X).$$

Now fix any $t \geq 0$ and take any rational $q > 0$. Because $\Gamma^{-1}(t)$ and $\Gamma^{-1}(t + q)$ are non-negative $\mathcal{F}_\infty$-measurable random variables, there exist non-decreasing sequences X_n and Z_n of $\mathcal{F}_\infty$-measurable simple random variables such that

$$\lim_{n \to \infty} X_n = \Gamma^{-1}(t), \quad \lim_{n \to \infty} Z_n = \Gamma^{-1}(t + q).$$

Because X_n and Z_n are simple random variables and the paths of Γ, and so of F, are continuous, we have

$$\lim_{n \to \infty} \mathbb{E}(\mathbf{1}_{\{\tau \leq X_n\}}|\mathcal{F}_\infty) = \lim_{n \to \infty} F(X_n)$$
$$= F(\Gamma^{-1}(t)) = 1 - e^{-\Gamma(\Gamma^{-1}(t))} = 1 - e^{-t},$$
$$\lim_{n \to \infty} \mathbb{E}(\mathbf{1}_{\{\tau \leq Z_n\}}|\mathcal{F}_\infty) = \lim_{n \to \infty} F(Z_n)$$
$$= F(\Gamma^{-1}(t + q)) = 1 - e^{-\Gamma(\Gamma^{-1}(t+q))} = 1 - e^{-(t+q)}.$$

Because Γ has strictly increasing paths, so does Γ^{-1}, hence

$$\Gamma^{-1}(t) < \Gamma^{-1}(t + q).$$

It follows that

$$\bigcup_{n=1}^{\infty} \{\tau \leq X_n\} \subset \{\tau \leq \Gamma^{-1}(t)\} \subset \bigcup_{n=1}^{\infty} \{\tau \leq Z_n\}.$$

As a result,

$$\lim_{n \to \infty} \mathbf{1}_{\{\tau \leq X_n\}} = \mathbf{1}_{\bigcup_{n=1}^{\infty}\{\tau \leq X_n\}} \leq \mathbf{1}_{\{\tau \leq \Gamma^{-1}(t)\}} \leq \mathbf{1}_{\bigcup_{n=1}^{\infty}\{\tau \leq Z_n\}} = \lim_{n \to \infty} \mathbf{1}_{\{\tau \leq Z_n\}},$$

and so

$$\lim_{n \to \infty} \mathbb{E}(\mathbf{1}_{\{\tau \leq X_n\}}|\mathcal{F}_\infty) \leq \mathbb{E}(\mathbf{1}_{\{\tau \leq \Gamma^{-1}(t)\}}|\mathcal{F}_\infty) \leq \lim_{n \to \infty} \mathbb{E}(\mathbf{1}_{\{\tau \leq Z_n\}}|\mathcal{F}_\infty).$$

This gives

$$1 - e^{-t} \leq \mathbb{E}(\mathbf{1}_{\{\tau \leq \Gamma^{-1}(t)\}}|\mathcal{F}_\infty) \leq 1 - e^{-(t+q)}$$

for every rational $q > 0$, hence

$$\mathbb{E}(\mathbf{1}_{\{\tau \leq \Gamma^{-1}(t)\}}|\mathcal{F}_\infty) = 1 - e^{-t}.$$

Observe that $\{\Gamma(\tau) \leq t\} = \{\tau \leq \Gamma^{-1}(t)\}$, which means that

$$\mathbb{E}(\mathbf{1}_{\{\Gamma(\tau) \leq t\}}|\mathcal{F}_\infty) = 1 - e^{-t}.$$

By taking the expectation on both sides of this equality, we obtain

$$Q(\Gamma(\tau) \leq t) = 1 - e^{-t}$$

for each $t \geq 0$, so $Y = \Gamma(\tau)$ does indeed have the unit exponential distribution. Moreover, $\mathbb{E}(\mathbf{1}_{\{\Gamma(\tau) \leq t\}} | \mathcal{F}_\infty) = 1 - e^{-t}$ is deterministic, so by Lemma 5.25 we find that $Y = \Gamma(\tau)$ and $\mathcal{F}_\infty$ are independent. □

5.5 Conditional expectations

We are going to explore the possibility of representing $(\mathcal{G}_t)_{t \geq 0}$-adapted processes in terms of $(\mathcal{F}_t)_{t \geq 0}$-adapted ones, and find expressions for conditional expectations with respect to $\mathcal{G}_t$ in terms of conditional expectations with respect to $\mathcal{F}_t$. In Chapter 2 we established that an $\mathcal{I}_t$-measurable random variable is constant on $\{t < \tau\}$. By analogy, in the current setting we are going to show that any $\mathcal{G}_t$-measurable random variable is equal to an $\mathcal{F}_t$-measurable random variable on $\{t < \tau\}$.

Lemma 5.27
For each $A \in \mathcal{G}_t$ there exists a $B \in \mathcal{F}_t$ such that

$$A \cap \{t < \tau\} = B \cap \{t < \tau\}.$$

Proof   Consider the family of events

$$\mathcal{A} = \{A \in \mathcal{G}_t : \exists B \in \mathcal{F}_t \text{ such that } A \cap \{t < \tau\} = B \cap \{t < \tau\}\}.$$

We shall show that $\mathcal{A}$ is a σ-field. When $A = \emptyset$, we take $B = \emptyset$. If $A = \Omega$, then $B = \Omega$ clearly does the job. Hence $\emptyset, \Omega \in \mathcal{A}$. If $A \in \mathcal{A}$, with a suitable $B \in \mathcal{F}_t$, then

$$\begin{aligned}
(\Omega \setminus A) \cap \{t < \tau\} &= \{t < \tau\} \setminus (A \cap \{t < \tau\}) \\
&= \{t < \tau\} \setminus (B \cap \{t < \tau\}) \\
&= (\Omega \setminus B) \cap \{t < \tau\},
\end{aligned}$$

so $\Omega \setminus A \in \mathcal{A}$. If $A_n \in \mathcal{A}$ with a corresponding $B_n \in \mathcal{F}_t$ for each $n = 1, 2, \ldots$, then

$$\begin{aligned}
\left(\bigcup_{n=1}^{\infty} A_n \right) \cap \{t < \tau\} &= \bigcup_{n=1}^{\infty} (A_n \cap \{t < \tau\}) \\
&= \bigcup_{n=1}^{n} (B_n \cap \{t < \tau\}) \\
&= \left(\bigcup_{n=1}^{\infty} B_n \right) \cap \{t < \tau\},
\end{aligned}$$

which means that $\bigcup_{n=1}^{\infty} A_n \in \mathcal{A}$.

To prove the claim of the lemma, which reads $\mathcal{G}_t = \mathcal{A}$, it is enough to show that $\mathcal{I}_t \subset \mathcal{A}$ and $\mathcal{F}_t \subset \mathcal{A}$. For the first inclusion it is sufficient to see that the generators of $\mathcal{I}_t$, namely the events $\{\tau \leq s\}$, belong to $\mathcal{A}$ for all $s \in [0, t]$. Clearly, $\{\tau \leq s\} \cap \{t < \tau\} = \varnothing \cap \{t < \tau\}$ and $B = \varnothing$ is good for each $s \in [0, t]$. To show the other inclusion, let $A \in \mathcal{F}_t$ and take $B = A$. $\quad\square$

Proposition 5.28

If X is $\mathcal{G}_t$-measurable, then there exists an $\mathcal{F}_t$-measurable random variable Y such that

$$X\mathbf{1}_{\{t<\tau\}} = Y\mathbf{1}_{\{t<\tau\}}.$$

Proof First take the indicator function $X = \mathbf{1}_A$ for some $A \in \mathcal{G}_t$. By Lemma 5.27, there exists a $B \in \mathcal{F}_t$ such that

$$A \cap \{t < \tau\} = B \cap \{t < \tau\},$$

so we can take $Y = \mathbf{1}_B$. Next, consider a step function X and construct Y by linearity. Finally, a general X can be expressed as the limit of a sequence of step functions X_n. For each n we take the corresponding $\mathcal{F}_t$-measurable random variable Y_n, and then

$$Y = \limsup_{n \to \infty} Y_n.$$

The equality $X_n \mathbf{1}_{\{t<\tau\}} = Y_n \mathbf{1}_{\{t<\tau\}}$ gives $X\mathbf{1}_{\{t<\tau\}} = Y\mathbf{1}_{\{t<\tau\}}$ in the limit. $\quad\square$

We are ready to find the conditional expectation with respect to $\mathcal{G}_t$ of the restriction $X\mathbf{1}_{\{t<\tau\}}$ of a random variable X to the pre-default event $\{t < \tau\}$. Recall the corresponding formula

$$\mathbb{E}(X\mathbf{1}_{\{t<\tau\}}|\mathcal{I}_t) = \mathbf{1}_{\{t<\tau\}} \frac{\mathbb{E}(X\mathbf{1}_{\{t<\tau\}})}{Q(t < \tau)}$$

proved in Proposition 2.24. It follows as a special case from the next result when $\mathcal{F}_t = \{\Omega, \varnothing\}$ and $\mathcal{G}_t = \mathcal{I}_t$.

Theorem 5.29

If X is integrable, then

$$\mathbb{E}(X\mathbf{1}_{\{t<\tau\}}|\mathcal{G}_t) = \mathbf{1}_{\{t<\tau\}} \frac{\mathbb{E}(X\mathbf{1}_{\{t<\tau\}}|\mathcal{F}_t)}{Q(t < \tau|\mathcal{F}_t)}.$$

Proof The random variable $\mathbb{E}(X\mathbf{1}_{\{t<\tau\}}|\mathcal{G}_t)$ is $\mathcal{G}_t$-measurable. By Proposition 5.28, there exists an $\mathcal{F}_t$-measurable random variable Y such that

$$\mathbb{E}(X\mathbf{1}_{\{t<\tau\}}|\mathcal{G}_t)\mathbf{1}_{\{t<\tau\}} = Y\mathbf{1}_{\{t<\tau\}}.$$

But $\{t < \tau\} \in \mathcal{I}_t \subset \mathcal{G}_t$, so the left-hand side simplifies, and

$$\mathbb{E}(X\mathbf{1}_{\{t<\tau\}}|\mathcal{G}_t) = Y\mathbf{1}_{\{t<\tau\}}. \tag{5.8}$$

By taking the conditional expectation with respect to $\mathcal{F}_t$ on both sides, we obtain

$$\mathbb{E}(\mathbb{E}(X\mathbf{1}_{\{t<\tau\}}|\mathcal{G}_t)|\mathcal{F}_t) = \mathbb{E}(Y\mathbf{1}_{\{t<\tau\}}|\mathcal{F}_t).$$

On the left we use the tower property to get $\mathbb{E}(X\mathbf{1}_{\{t<\tau\}}|\mathcal{F}_t)$ since $\mathcal{F}_t \subset \mathcal{G}_t$. On the right we employ the measurability of Y with respect to $\mathcal{F}_t$, which gives $Y\mathbb{E}(\mathbf{1}_{\{t<\tau\}}|\mathcal{F}_t)$. We can put these together to get

$$\mathbb{E}(X\mathbf{1}_{\{t<\tau\}}|\mathcal{F}_t) = Y\mathbb{E}(\mathbf{1}_{\{t<\tau\}}|\mathcal{F}_t) = YQ(t < \tau|\mathcal{F}_t).$$

From this we evaluate Y and insert it in (5.8) to complete the proof. $\square$

It remains to find out what happens on $\{\tau \leq t\}$, that is, when a default takes place before or at time t. We shall see that to find the conditional expectation of $X\mathbf{1}_{\{\tau\leq t\}}$ with respect to $\mathcal{G}_t$, conditioning with respect to the smallest σ-field containing $\sigma(\tau)$ and $\mathcal{F}_t$ is needed. We denote this σ-field by

$$\mathcal{H}_t = \sigma(\sigma(\tau) \cup \mathcal{F}_t).$$

Since $\mathcal{I}_t \subset \sigma(\tau)$, it follows that $\mathcal{G}_t \subset \mathcal{H}_t$. Let us begin with an auxiliary lemma.

Lemma 5.30
If $A \in \mathcal{H}_t$, then $A \cap \{\tau \leq t\} \in \mathcal{G}_t$.

Proof Let

$$\mathcal{A} = \{A \in \mathcal{H}_t : A \cap \{\tau \leq t\} \in \mathcal{G}_t\}.$$

We need to show that $\mathcal{A}$ is a σ-field containing $\sigma(\tau)$ and $\mathcal{F}_t$. This is done in Exercise 5.8. It follows that $\mathcal{A} = \mathcal{H}_t$, which proves the lemma. $\square$

Exercise 5.8 Show that $\mathcal{A}$ is a σ-field containing $\sigma(\tau)$ and $\mathcal{F}_t$.

The next step is to extend this relationship to random variables.

Lemma 5.31
For any random variable X, if $X\mathbf{1}_{\{\tau\leq t\}}$ is $\mathcal{H}_t$-measurable, then $X\mathbf{1}_{\{\tau\leq t\}}$ is $\mathcal{G}_t$-measurable.

Proof If $X = \mathbf{1}_A$, then $\mathcal{H}_t$-measurability of $X\mathbf{1}_{\{\tau \le t\}}$ implies that $A \cap \{\tau \le t\} \in \mathcal{H}_t$, and Lemma 5.30 gives the desired conclusion. This, then, routinely extends to simple (step) random variables, and passing to a limit gives the result for arbitrary random variables. □

The next result is an immediate consequence of this lemma.

Theorem 5.32
If X is integrable, then

$$\mathbb{E}(X\mathbf{1}_{\{\tau \le t\}}|\mathcal{G}_t) = \mathbb{E}(X\mathbf{1}_{\{\tau \le t\}}|\mathcal{H}_t).$$

Proof Since $\{\tau \le t\} \in \sigma(\tau)$, it follows that

$$\mathbb{E}(X\mathbf{1}_{\{\tau \le t\}}|\mathcal{H}_t) = \mathbb{E}(X|\mathcal{H}_t)\mathbf{1}_{\{\tau \le t\}},$$

so $\mathbb{E}(X\mathbf{1}_{\{\tau \le t\}}|\mathcal{H}_t)$ is $\mathcal{G}_t$-measurable by Lemma 5.31. By the definition of conditional expectation,

$$\mathbb{E}(\mathbf{1}_A X\mathbf{1}_{\{\tau \le t\}}) = \mathbb{E}(\mathbf{1}_A \mathbb{E}(X\mathbf{1}_{\{\tau \le t\}}|\mathcal{H}_t))$$

for each $A \in \mathcal{H}_t$, and thus for each $A \in \mathcal{G}_t$ given that $\mathcal{G}_t \subset \mathcal{H}_t$. Because $\mathbb{E}(X\mathbf{1}_{\{\tau \le t\}}|\mathcal{H}_t)$ is $\mathcal{G}_t$-measurable, we apply the definition of conditional expectation once again to get

$$\mathbb{E}(X\mathbf{1}_{\{\tau \le t\}}|\mathcal{G}_t) = \mathbb{E}(X\mathbf{1}_{\{\tau \le t\}}|\mathcal{H}_t),$$

completing the proof. □

Theorem 5.29 together with Theorem 5.32 lead to the following conclusion.

Corollary 5.33
If X is integrable, then

$$\mathbb{E}(X|\mathcal{G}_t) = \mathbf{1}_{\{\tau \le t\}}\mathbb{E}(X|\mathcal{H}_t) + \mathbf{1}_{\{t < \tau\}}e^{\Gamma(t)}\mathbb{E}(X\mathbf{1}_{\{t < \tau\}}|\mathcal{F}_t).$$

We can obtain a more explicit expression on the right-hand side of this equality when X has a special form, which will be sufficient for our purposes.

Theorem 5.34
Consider a process $h(t), t \ge 0$, adapted to the filtration $(\mathcal{F}_t)_{t \ge 0}$, with left-continuous paths and such that $h(\tau)$ is integrable. Then for any $s \ge 0$

$$\mathbb{E}(h(\tau)|\mathcal{G}_s) = \mathbf{1}_{\{\tau \le s\}}h(\tau) + \mathbf{1}_{\{s < \tau\}}e^{\Gamma(s)}\mathbb{E}\left(\int_s^\infty h(u)f(u)du \middle| \mathcal{F}_s\right).$$

Proof   By Corollary 5.33,

$$\mathbb{E}(h(\tau)|\mathcal{G}_s) = \mathbf{1}_{\{\tau \le s\}}\mathbb{E}(h(\tau)|\mathcal{H}_s) + \mathbf{1}_{\{s < \tau\}}e^{\Gamma(s)}\mathbb{E}(h(\tau)\mathbf{1}_{\{s < \tau\}}|\mathcal{F}_s).$$

We shall show that the two terms on the right-hand side match the corresponding terms in the claim of the theorem.

We start with the case when $s > 0$. Because h has left-continuous paths, we have, almost surely,

$$h(t) = \lim_{n \to \infty} h_n(t) \quad \text{for each } t > 0, \tag{5.9}$$

where

$$h_n(t) = \sum_{i=1}^{n^2} h(s_n^{i-1})\mathbf{1}_{(s_n^{i-1}, s_n^i]}(t)$$

with $s_n^i = \frac{is}{n}$ for each $n, i = 1, 2, \ldots$ (see Exercise 5.9). Observe that $h(\tau)\mathbf{1}_{\{\tau \le s\}}$ is $\mathcal{H}_s$-measurable since, almost surely,

$$h(\tau)\mathbf{1}_{\{\tau \le s\}} = \lim_{n \to \infty} h_n(\tau)\mathbf{1}_{\{\tau \le s\}} = \lim_{n \to \infty} \sum_{i=1}^{n} h(s_n^{i-1})\mathbf{1}_{(s_n^{i-1}, s_n^i]}(\tau),$$

where $\mathbf{1}_{(s_n^{i-1}, s_n^i]}(\tau)$ is $\sigma(\tau)$-measurable and $h(s_n^{i-1})$ is $\mathcal{F}_{s_n^{i-1}}$-measurable, hence $\mathcal{F}_s$-measurable, for every $n = 1, 2, \ldots$ and every $i = 1, \ldots, n$. It follows that

$$\mathbf{1}_{\{\tau \le s\}}\mathbb{E}(h(\tau)|\mathcal{H}_s) = \mathbb{E}(\mathbf{1}_{\{\tau \le s\}}h(\tau)|\mathcal{H}_s) = \mathbf{1}_{\{\tau \le s\}}h(\tau).$$

It remains to show that

$$\mathbb{E}(h(\tau)\mathbf{1}_{\{s < \tau\}}|\mathcal{F}_s) = \mathbb{E}\left(\left.\int_s^\infty h(u)f(u)du\right|\mathcal{F}_s\right).$$

First we verify this equality for $h_n(t)$. Indeed,

$$\mathbb{E}(h_n(\tau)\mathbf{1}_{\{s < \tau\}}|\mathcal{F}_s) = \sum_{i=n+1}^{n^2} \mathbb{E}(h(s_n^{i-1})\mathbf{1}_{(s_n^{i-1}, s_n^i]}(\tau)|\mathcal{F}_s)$$

$$= \sum_{i=n+1}^{n^2} \mathbb{E}(\mathbb{E}(h(s_n^{i-1})\mathbf{1}_{(s_n^{i-1}, s_n^i]}(\tau)|\mathcal{F}_{s_n^i})|\mathcal{F}_s)$$

$$= \sum_{i=n+1}^{n^2} \mathbb{E}(h(s_n^{i-1})\mathbb{E}(\mathbf{1}_{(s_n^{i-1}, s_n^i]}(\tau)|\mathcal{F}_{s_n^i})|\mathcal{F}_s)$$

$$= \sum_{i=n+1}^{n^2} \mathbb{E}(h(s_n^{i-1})(\mathbb{E}(\mathbf{1}_{\{\tau \le s_n^i\}}|\mathcal{F}_{s_n^i}) - \mathbb{E}(\mathbf{1}_{\{\tau \le s_n^{i-1}\}}|\mathcal{F}_{s_n^i}))|\mathcal{F}_s).$$

We use Lemma 5.15 to get

$$\mathbb{E}(h_n(\tau)\mathbf{1}_{\{s<\tau\}}|\mathcal{F}_s) = \sum_{i=n+1}^{n^2} \mathbb{E}(h(s_n^{i-1})(F(s_n^i) - F(s_n^{i-1}))|\mathcal{F}_s)$$

$$= \mathbb{E}\left(\int_s^\infty h_n(u)f(u)du \,\middle|\, \mathcal{F}_s\right).$$

It follows from (5.9) that, almost surely,

$$h(\tau)\mathbf{1}_{\{s<\tau\}} = \lim_{n\to\infty} h_n(\tau)\mathbf{1}_{\{s<\tau\}}.$$

Now suppose that h is bounded. It follows by dominated convergence that

$$\mathbb{E}(h(\tau)\mathbf{1}_{\{s<\tau\}}|\mathcal{F}_s) = \lim_{n\to\infty} \mathbb{E}(h_n(\tau)\mathbf{1}_{\{s<\tau\}}|\mathcal{F}_s)$$

$$= \lim_{n\to\infty} \mathbb{E}\left(\int_s^\infty h_n(u)f(u)du \,\middle|\, \mathcal{F}_s\right)$$

$$= \mathbb{E}\left(\int_s^\infty h(u)f(u)du \,\middle|\, \mathcal{F}_s\right).$$

If h is non-negative (but not necessarily bounded), then it can be approximated by the non-decreasing sequence $\min(n, h)$ of bounded adapted processes with left-continuous paths, and $h(\tau)$ can be approximated by the non-decreasing sequence $\min(n, h(\tau))$. Hence monotone convergence gives the desired formula for any such h. Finally, for a general process h satisfying the assumptions of this theorem, we can split h into positive and negative parts h^+ and h^-, both of which have been shown to satisfy the claimed equality. The integrability of $h(\tau)$ ensures that $h^+(\tau)$ and $h^-(\tau)$ are integrable, hence we can assemble the corresponding equalities to obtain the desired formula for h. This completes the proof in the case when $s > 0$. For $s = 0$, see Exercise 5.10. □

Exercise 5.9 Let h be a left-continuous function from $[0, \infty)$ to $\mathbb{R}$ and let $s > 0$. Show that for each $t > 0$

$$h(t) = \lim_{n\to\infty} h_n(t),$$

where $h_n(t) = \sum_{i=1}^{n^2} h(s_n^{i-1})\mathbf{1}_{(s_n^{i-1}, s_n^i]}(t)$ with $s_n^i = \frac{is}{n}$ for each $n, i = 1, 2, \ldots$.

Exercise 5.10   Complete the proof of Theorem 5.34 in the case when $s = 0$.

The following result will provide us with a formula for defaultable security prices. To compute the quantity on the right we need: the filtration $(\mathcal{F}_t)_{t \geq 0}$, assumed to be known, the form of the hazard process Γ, which is a matter of some modelling assumptions, and a random variable Y, which will represent the time T payoff of a defaultable security.

Theorem 5.35
For any $0 \leq t \leq T$, if Y is integrable and $\mathcal{F}_T$-measurable, then

$$\mathbb{E}(Y\mathbf{1}_{\{T<\tau\}}|\mathcal{G}_t) = \mathbf{1}_{\{t<\tau\}}\mathbb{E}(Ye^{-(\Gamma(T)-\Gamma(t))}|\mathcal{F}_t).$$

Proof   We can transform the left-hand side of the equality as follows:

$$
\begin{aligned}
\mathbb{E}(Y\mathbf{1}_{\{T<\tau\}}|\mathcal{G}_t) &= \mathbb{E}(Y\mathbf{1}_{\{T<\tau\}}\mathbf{1}_{\{t<\tau\}}|\mathcal{G}_t) & \text{since } t \leq T\\
&= \mathbf{1}_{\{t<\tau\}}e^{\Gamma(t)}\mathbb{E}(Y\mathbf{1}_{\{T<\tau\}}\mathbf{1}_{\{t<\tau\}}|\mathcal{F}_t) & \text{by Theorem 5.29}\\
&= \mathbf{1}_{\{t<\tau\}}e^{\Gamma(t)}\mathbb{E}(Y\mathbf{1}_{\{T<\tau\}}|\mathcal{F}_t) & \text{since } t \leq T\\
&= \mathbf{1}_{\{t<\tau\}}e^{\Gamma(t)}\mathbb{E}(\mathbb{E}(Y\mathbf{1}_{\{T<\tau\}}|\mathcal{F}_T)|\mathcal{F}_t) & \text{by the tower property}\\
&= \mathbf{1}_{\{t<\tau\}}e^{\Gamma(t)}\mathbb{E}(Y\mathbb{E}(\mathbf{1}_{\{T<\tau\}}|\mathcal{F}_T)|\mathcal{F}_t) & \text{since } Y \text{ is } \mathcal{F}_T\text{-measurable}\\
&= \mathbf{1}_{\{t<\tau\}}e^{\Gamma(t)}\mathbb{E}(Ye^{-\Gamma(T)}|\mathcal{F}_t) & \text{by the definition of } \Gamma(T)\\
&= \mathbf{1}_{\{t<\tau\}}\mathbb{E}(Ye^{-(\Gamma(T)-\Gamma(t))}|\mathcal{F}_t) & \text{since } \Gamma(t) \text{ is } \mathcal{F}_t\text{-measurable.}
\end{aligned}
$$

This completes the argument.   □

5.6   Martingale properties

We are going to generalise the results in Sections 3.4 and 3.5 concerned with the martingales associated with the default indicator process. Just like in Section 3.4, we say that $\Gamma(t \wedge \tau)$ is a **compensator** of the indicator process $I(t)$, given that Theorem 3.15 can be extended as follows.

Theorem 5.36
*The **compensated indicator** process*

$$M(t) = I(t) - \Gamma(t \wedge \tau)$$

is a martingale with respect to $(\mathcal{G}_t)_{t \geq 0}$.

Proof First of all, we need to show that $M(t)$ is integrable. Clearly, $I(t)$ is integrable as it is bounded, so we need to show the integrability of $\Gamma(t \wedge \tau)$. By Theorem 5.34,

$$\mathbb{E}(\Gamma(t \wedge \tau)) = \mathbb{E}\left(\int_0^{t \wedge \tau} \gamma(u)du \right)$$

$$= \mathbb{E}\left(\int_0^\infty \left(\int_0^{t \wedge v} \gamma(u)du \right) f(v)dv \right).$$

By changing the order of integration, we obtain

$$\mathbb{E}(\Gamma(t \wedge \tau)) = \mathbb{E}\left(\int_0^t \gamma(u) \left(\int_u^\infty f(v)dv \right) du \right)$$

$$= \mathbb{E}\left(\int_0^t \gamma(u)G(u)du \right)$$

$$= \mathbb{E}\left(\int_0^t f(u)du \right) = \mathbb{E}\,(F(t)) \le 1.$$

Next, the goal is to show that for any $0 \le s \le t$

$$\mathbb{E}(I(t) - I(s)|\mathcal{G}_s) = \mathbb{E}(\Gamma(t \wedge \tau) - \Gamma(s \wedge \tau)|\mathcal{G}_s). \tag{5.10}$$

On the left, using Theorem 5.35, we have

$$\mathbb{E}(I(t) - I(s)|\mathcal{G}_s) = \mathbb{E}(\mathbf{1}_{\{\tau \le t\}}|\mathcal{G}_s) - \mathbf{1}_{\{\tau \le s\}}$$

$$= 1 - \mathbb{E}(\mathbf{1}_{\{t < \tau\}}|\mathcal{G}_s) - \mathbf{1}_{\{\tau \le s\}}$$

$$= \mathbf{1}_{\{s < \tau\}} - \mathbf{1}_{\{s < \tau\}} e^{\Gamma(s)} \mathbb{E}(e^{-\Gamma(t)}|\mathcal{F}_s).$$

The right-hand side of (5.10) vanishes on $\{\tau \le s\}$ since

$$\mathbf{1}_{\{\tau \le s\}}(\Gamma(t \wedge \tau) - \Gamma(s \wedge \tau)) = 0,$$

so it has the form

$$\mathbb{E}(\Gamma(t \wedge \tau) - \Gamma(s \wedge \tau)|\mathcal{G}_s) = \mathbb{E}(\mathbf{1}_{\{s < \tau\}}(\Gamma(t \wedge \tau) - \Gamma(s))|\mathcal{G}_s).$$

Let us use Theorem 5.34 with $h(u) = \mathbf{1}_{\{s < u\}}(\Gamma(t \wedge u) - \Gamma(s))$, which is adapted to $(\mathcal{F}_t)_{t \ge 0}$ and left-continuous since Γ is continuous by Assumption 5.18. This gives

$$\mathbb{E}(\Gamma(t \wedge \tau) - \Gamma(s \wedge \tau)|\mathcal{G}_s)$$

$$= \mathbf{1}_{\{s < \tau\}} e^{\Gamma(s)} \mathbb{E}\left(\int_s^\infty \mathbf{1}_{\{s < u\}}(\Gamma(t \wedge u) - \Gamma(s)) f(u)du \,\Big|\, \mathcal{F}_s \right)$$

$$= \mathbf{1}_{\{s < \tau\}} e^{\Gamma(s)} \mathbb{E}\left(\int_s^\infty (\Gamma(t \wedge u) - \Gamma(s)) f(u)du \,\Big|\, \mathcal{F}_s \right).$$

We integrate by parts to compute the last integral:

$$\int_s^\infty (\Gamma(t \wedge u) - \Gamma(s))f(u)du$$

$$= \int_s^t \Gamma(u)f(u)du + \int_t^\infty \Gamma(t)f(u)du - \int_s^\infty \Gamma(s)f(u)du$$

$$= \int_s^t \Gamma(u)f(u)du + \Gamma(t)(1 - F(t)) - \Gamma(s)(1 - F(s))$$

$$= -\int_s^t \gamma(u)F(u)du + \Gamma(t) - \Gamma(s)$$

$$= -\int_s^t (1 - e^{-\Gamma(u)})\gamma(u)du + \Gamma(t) - \Gamma(s)$$

$$= \int_s^t e^{-\Gamma(u)}\gamma(u)du$$

$$= e^{-\Gamma(s)} - e^{-\Gamma(t)}.$$

Therefore, the right-hand side of (5.10) takes the same form as the left-hand side, namely

$$\mathbf{1}_{\{s<\tau\}}e^{\Gamma(s)}\mathbb{E}(e^{-\Gamma(s)} - e^{-\Gamma(t)}|\mathcal{F}_s) = \mathbf{1}_{\{s<\tau\}} - \mathbf{1}_{\{s<\tau\}}e^{\Gamma(s)}\mathbb{E}(e^{-\Gamma(t)}|\mathcal{F}_s).$$

$$\square$$

The next exercise shows that Proposition 3.17 can also be extended to the current setting.

Exercise 5.11   Show that the process

$$L(t) = (1 - I(t))e^{\Gamma(t)}$$

is a $(\mathcal{G}_t)_{t\geq0}$-martingale.

Remark 5.37

An alternative short proof of Theorem 5.36 can be presented using the theory of stochastic integration with respect to martingales outlined in Appendix A.3, together with the fact, shown in Exercise 5.11, that $L(t) = (1 - I(t))e^{\Gamma(t)}$ is a martingale with respect to the filtration $(\mathcal{G}_t)_{t\geq0}$. This is because the compensated indicator process M can be represented as an integral with respect to L, namely

$$M(t) = I(t) - \Gamma(t \wedge \tau)$$

$$= \mathbf{1}_{\{\tau \le t\}} - \int_0^{t \wedge \tau} \gamma(u) du$$

$$= - \int_0^t e^{-\Gamma(u)} e^{\Gamma(u)} d(1 - I(u)) - \int_0^t e^{-\Gamma(u)} (1 - I(u)) d(e^{\Gamma(u)})$$

$$= - \int_0^t e^{-\Gamma(u)} dL(u).$$

Here $e^{-\Gamma(u)}$ is bounded, so the integral with respect to L is a $(\mathcal{G}_t)_{t \ge 0}$-martingale because L is a $(\mathcal{G}_t)_{t \ge 0}$-martingale.

Exercise 5.12 Show that if $X(t)$ is a martingale with respect to the filtration $(\mathcal{F}_t)_{t \ge 0}$, then $X(t)L(t)$ is a martingale with respect to $(\mathcal{G}_t)_{t \ge 0}$.

Because $M(t) = I(t) - \Gamma(t \wedge \tau)$ is the difference of two processes with non-decreasing paths, just as in Section 3.5, the integral $\int_0^t X(u) dM(u)$ of a process X with respect to M can be understood as a **pathwise** Lebesgue–Stieltjes integral. Moreover, formula (3.4) also extends to the present case,

$$\int_0^t X(u) dM(u) = \int_0^t X(u) dI(u) - \int_0^t X(u) d\Gamma(u \wedge \tau)$$

$$= \mathbf{1}_{\{\tau \le t\}} X(\tau) - \int_0^{t \wedge \tau} X(u) \gamma(u) du. \qquad (5.11)$$

The following lemma will be needed in the remainder of this section.

Lemma 5.38
Suppose that $X(t), t \ge 0$, is an $(\mathcal{F}_t)_{t \ge 0}$-adapted process with left-continuous paths. If $X(\tau)$ is integrable, then so is $\int_0^t X(u) dM(u)$ for each $t \ge 0$.

Proof The first term on the right-hand side of (5.11) is clearly integrable when $X(\tau)$ is integrable, so we just need to show integrability for the other term. Applying Theorem 5.34, we get

$$\mathbb{E} \left| \int_0^{t \wedge \tau} X(u) \gamma(u) du \right| \le \mathbb{E} \left(\int_0^{t \wedge \tau} |X(u)| \gamma(u) du \right)$$

$$= \mathbb{E} \left(\int_0^\infty \left(\int_0^{t \wedge v} |X(u)| \gamma(u) du \right) f(v) dv \right).$$

Changing the order of integration, we then have

$$\mathbb{E}\left|\int_0^{t\wedge\tau} X(u)\gamma(u)du\right| \leq \mathbb{E}\left(\int_0^t |X(u)|\,\gamma(u)\left(\int_u^\infty f(v)dv\right)du\right)$$

$$= \mathbb{E}\left(\int_0^t |X(u)|\,\gamma(u)G(u)du\right)$$

$$= \mathbb{E}\left(\int_0^t |X(u)|\,f(u)du\right)$$

$$= \mathbb{E}\left(\mathbf{1}_{\{\tau\leq t\}}\,|X(\tau)|\right)$$

$$\leq \mathbb{E}\,|X(\tau)| < \infty,$$

where we use Theorem 5.34 once again in the last equality.   □

We are going to show that, as one might expect, Theorem 3.20 can be extended to the current setting. We begin with a generalisation of Lemma 3.19, with a deterministic function X replaced by a process adapted to the filtration $(\mathcal{F}_t)_{t\geq0}$ and having left-continuous paths.

Theorem 5.39
If X is an $(\mathcal{F}_t)_{t\geq0}$-adapted process with left-continuous paths such that $X(\tau)$ is integrable and the Lebesgue–Stieltjes integral $\int_0^t X(u)dM(u)$ exists for all $t \geq 0$, then the process $\int_0^t X(u)dM(u)$ is a martingale with respect to the filtration $(\mathcal{G}_t)_{t\geq0}$.

Proof   By formula (5.11), we can write the integral as

$$\int_0^t X(u)dM(u) = h_t(\tau),$$

where

$$h_t(v) = \mathbf{1}_{\{v\leq t\}}X(v) - \int_0^{t\wedge v} X(u)\gamma(u)du.$$

According to Lemma 5.38, $h_t(\tau)$ is integrable since $X(\tau)$ is. Because X has left-continuous paths and is $(\mathcal{F}_t)_{t\geq0}$-adapted, it follows that $h_t(v)$ considered as a function of v has left-continuous paths and is $\mathcal{F}_v$-measurable for each

$v \geq 0$. Hence we can apply Theorem 5.34 to find that, for any $0 \leq s \leq t$,

$$\mathbb{E}\left(\int_0^t X(u)dM(u)\,\Big|\,\mathcal{G}_s\right)$$

$$= \mathbb{E}(h_t(\tau)|\mathcal{G}_s)$$

$$= \mathbf{1}_{\{\tau \leq s\}}h_t(\tau) + \mathbf{1}_{\{s<\tau\}}e^{\Gamma(s)}\mathbb{E}\left(\int_s^\infty h_t(u)f(u)du\,\Big|\,\mathcal{F}_s\right)$$

$$= \mathbf{1}_{\{\tau \leq s\}}X(\tau) - \mathbf{1}_{\{\tau \leq s\}}\int_0^\tau X(u)\gamma(u)du$$

$$\quad + \mathbf{1}_{\{s<\tau\}}e^{\Gamma(s)}\mathbb{E}\left(\int_s^t X(u)f(u)du\,\Big|\,\mathcal{F}_s\right)$$

$$\quad - \mathbf{1}_{\{s<\tau\}}e^{\Gamma(s)}\mathbb{E}\left(\int_s^\infty \left(\int_0^{t\wedge u} X(v)\gamma(v)dv\right)f(u)du\,\Big|\,\mathcal{F}_s\right).$$

The last expectation can be expressed as

$$\mathbb{E}\left(\int_s^\infty \left(\int_0^{t\wedge u} X(v)\gamma(v)dv\right)f(u)du\,\Big|\,\mathcal{F}_s\right)$$

$$= \mathbb{E}\left(\int_s^\infty \left(\int_0^s X(v)\gamma(v)dv\right)f(u)du\,\Big|\,\mathcal{F}_s\right)$$

$$\quad + \mathbb{E}\left(\int_s^\infty \left(\int_s^{t\wedge u} X(v)\gamma(v)dv\right)f(u)du\,\Big|\,\mathcal{F}_s\right)$$

$$= \mathbb{E}\left(\int_s^\infty f(u)du \int_0^s X(v)\gamma(v)dv\,\Big|\,\mathcal{F}_s\right)$$

$$\quad + \mathbb{E}\left(\int_s^t X(v)\gamma(v)\left(\int_v^\infty f(u)du\right)dv\,\Big|\,\mathcal{F}_s\right)$$

$$= \mathbb{E}\left(G(s)\int_0^s X(v)\gamma(v)dv\,\Big|\,\mathcal{F}_s\right) + \mathbb{E}\left(\int_s^t X(v)\gamma(v)G(v)dv\,\Big|\,\mathcal{F}_s\right)$$

$$= G(s)\int_0^s X(v)\gamma(v)dv + \mathbb{E}\left(\int_s^t X(v)f(v)dv\,\Big|\,\mathcal{F}_s\right).$$

On substituting this expression for the expectation, we see some cancellations, and get

$$\mathbb{E}\left(\int_0^t X(u)du \Big| \mathcal{G}_s\right)$$

$$= \mathbf{1}_{\{\tau \le s\}}X(\tau) - \mathbf{1}_{\{\tau \le s\}}\int_0^\tau X(u)\gamma(u)du - \mathbf{1}_{\{s < \tau\}}\int_0^s X(v)\gamma(v)dv$$

$$= \int_0^s X(u)dM(u).$$

$\square$

The next step is to extend the result from $(\mathcal{F}_t)_{t \ge 0}$-adapted to $(\mathcal{G}_t)_{t \ge 0}$-adapted processes X with left-continuous paths.

Theorem 5.40
If X is a process with left-continuous paths, adapted to the filtration $(\mathcal{G}_t)_{t \ge 0}$ and such that $X(\tau)$ is integrable and the integral $\int_0^t X(u)dM(u)$ exists for all $t \ge 0$, then the process $\int_0^t X(u)dM(u)$ is a $(\mathcal{G}_t)_{t \ge 0}$-martingale.

Proof First suppose that $|X| \le C$ for some constant $C > 0$. Because X is adapted to the filtration $(\mathcal{G}_t)_{t \ge 0}$, by Proposition 5.28, for each $t \ge 0$ there is an $\mathcal{F}_t$-measurable random variable $Y(t)$ such that

$$X(t)\mathbf{1}_{\{t < \tau\}} = Y(t)\mathbf{1}_{\{t < \tau\}}. \tag{5.12}$$

It follows that

$$Y(t) = e^{\Gamma(t)}\mathbb{E}(X(t)\mathbf{1}_{\{t < \tau\}}|\mathcal{F}_t),$$

so

$$|Y(t)| \le e^{\Gamma(t)}\mathbb{E}(|X(t)|\,\mathbf{1}_{\{t < \tau\}}|\mathcal{F}_t) \le Ce^{\Gamma(t)}\mathbb{E}(\mathbf{1}_{\{t < \tau\}}|\mathcal{F}_t) = C.$$

Hence, we have an $(\mathcal{F}_t)_{t \ge 0}$-adapted process Y such that (5.12) holds and $|Y| \le C$. Since X has left-continuous paths, it follows that each path of Y is left-continuous on $[0, \tau)$. Also, for each path of Y

$$\lim_{s \nearrow \tau} Y(s) = \lim_{s \nearrow \tau} X(s) = X(\tau).$$

We would like to apply Theorem 5.39 to Y, but there is no guarantee that Y is left-continuous outside the interval $[0, \tau)$. To work around this difficulty we approximate Y by the sequence

$$Y_n(s) = Y(0)\mathbf{1}_{\{0\}}(s) + \sum_{i=1}^{n^2} Y(t_n^{i-1})\mathbf{1}_{(t_n^{i-1}, t_n^i]}(s),$$

where $t_n^i = \frac{i}{n}$ for each $n = 1, 2, \dots$ and $i = 0, 1, \dots$. For each n, the process Y_n is adapted to the filtration $(\mathcal{F}_t)_{t\geq 0}$ and has left-continuous paths. Hence, by Theorem 5.39, $\int_0^t Y_n(u)dM(u)$ is a $(\mathcal{G}_t)_{t\geq 0}$-martingale. Moreover, $|Y_n| \leq C$ since $|Y| \leq C$. Because Y is left-continuous on $[0, \tau)$, we have

$$\lim_{n\to\infty} Y_n(t) = Y(t) = X(t) \quad \text{on } \{t < \tau\}.$$

Moreover, since $Y_n(\tau) = Y(t_n^{i-1}) = X(t_n^{i-1})$, where $t_n^{i-1} < \tau \leq t_n^i$, it follows that

$$\lim_{n\to\infty} Y_n(\tau) = \lim_{t\nearrow\tau} Y(t) = \lim_{t\nearrow\tau} X(t) = X(\tau).$$

As a result, for each $t \geq 0$

$$\lim_{n\to\infty} \int_0^t Y_n(u)dM(u) = \mathbf{1}_{\{\tau\leq t\}} \lim_{n\to\infty} Y_n(\tau) - \lim_{n\to\infty} \int_0^{t\wedge\tau} Y_n(u)\gamma(u)du$$

$$= \mathbf{1}_{\{\tau\leq t\}} X(\tau) - \int_0^{t\wedge\tau} X(u)\gamma(u)du$$

$$= \int_0^t X(u)dM(u),$$

where we use the dominated convergence theorem together with the fact that $|Y_n| \leq C$ for each n. Hence, using dominated convergence once again, we find that for any $0 \leq s \leq t$

$$\mathbb{E}\left(\int_0^t X(u)dM(u)\bigg| \mathcal{G}_s\right) = \lim_{n\to\infty} \mathbb{E}\left(\int_0^t Y_n(u)dM(u)\bigg| \mathcal{G}_s\right)$$

$$= \lim_{n\to\infty} \int_0^s Y_n(u)dM(u)$$

$$= \int_0^s X(u)dM(u)$$

because $\int_0^t Y_n(u)dM(u)$ is a $(\mathcal{G}_t)_{t\geq 0}$-martingale for each n. This shows that $\int_0^t X(u)dM(u)$ is also a $(\mathcal{G}_t)_{t\geq 0}$-martingale.

Next, we relax the condition that X is bounded and work under the assumption that X is non-negative. Consider the non-decreasing sequence X_n of non-negative bounded $(\mathcal{G}_t)_{t\geq 0}$-adapted processes with left-continuous paths defined as

$$X_n(t) = \min(n, X(t)).$$

It follows by monotone convergence that, almost surely,

$$\lim_{n\to\infty} \mathbf{1}_{\{\tau\leq s\}} X_n(\tau) = \mathbf{1}_{\{\tau\leq s\}} X(\tau),$$

$$\lim_{n\to\infty} \mathbb{E}(\mathbf{1}_{\{\tau\leq t\}} X_n(\tau)|\mathcal{G}_s) = \mathbb{E}(\mathbf{1}_{\{\tau\leq t\}} X(\tau)|\mathcal{G}_s),$$

and

$$\lim_{n\to\infty} \int_0^{s\wedge\tau} X_n(u)\gamma(u)du = \int_0^{s\wedge\tau} X(u)\gamma(u)du,$$

$$\lim_{n\to\infty} \mathbb{E}\left(\int_0^{t\wedge\tau} X_n(u)\gamma(u)du \middle| \mathcal{G}_s\right) = \mathbb{E}\left(\int_0^{t\wedge\tau} X(u)\gamma(u)du \middle| \mathcal{G}_s\right),$$

so

$$\lim_{n\to\infty} \int_0^{s\wedge\tau} X_n(u)dM(u) = \int_0^{s\wedge\tau} X(u)dM(u),$$

$$\lim_{n\to\infty} \mathbb{E}\left(\int_0^{t\wedge\tau} X_n(u)dM(u) \middle| \mathcal{G}_s\right) = \mathbb{E}\left(\int_0^{t\wedge\tau} X(u)dM(u) \middle| \mathcal{G}_s\right).$$

We can conclude that, almost surely,

$$\mathbb{E}\left(\int_0^{t\wedge\tau} X(u)dM(u) \middle| \mathcal{G}_s\right) = \lim_{n\to\infty} \mathbb{E}\left(\int_0^{t\wedge\tau} X_n(u)dM(u) \middle| \mathcal{G}_s\right)$$

$$= \lim_{n\to\infty} \int_0^{s\wedge\tau} X_n(u)dM(u)$$

$$= \int_0^{s\wedge\tau} X(u)dM(u).$$

If $X(\tau)$ is integrable in addition to X being non-negative, then

$$\mathbb{E}\left(\int_0^{t\wedge\tau} X(u)\gamma(u)du\right) = \lim_{n\to\infty} \mathbb{E}\left(\int_0^{t\wedge\tau} X_n(u)\gamma(u)du\right)$$

$$= \lim_{n\to\infty} \mathbb{E}\left(\mathbf{1}_{\{\tau\le t\}}X_n(\tau)\right) = \mathbb{E}\left(\mathbf{1}_{\{\tau\le t\}}X(\tau)\right) \le \mathbb{E}\left(X(\tau)\right) < \infty,$$

which means that

$$\int_0^t X(u)dM(u) = \mathbf{1}_{\{\tau\le t\}}X(\tau) - \int_0^{t\wedge\tau} X(u)\gamma(u)du$$

is integrable, so we can conclude that $\int_0^t X(u)dM(u)$ is a $(\mathcal{G}_t)_{t\ge0}$-martingale.

Finally, we take $X(\tau)$ to be integrable and relax the condition that X is non-negative. In this case, we consider the positive and negative parts X^+ and X^- of X, which have left-continuous paths and are $(\mathcal{G}_t)_{t\ge0}$-adapted, just like X. Moreover $X^+(\tau)$ and $X^-(\tau)$ are integrable because $X(\tau)$ is. It follows from the above that $\int_0^t X^+(u)dM(u)$ and $\int_0^t X^-(u)dM(u)$ are $(\mathcal{G}_t)_{t\ge0}$-martingales, which in turn implies that

$$\int_0^t X(u)dM(u) = \int_0^t X^+(u)dM(u) - \int_0^t X^-(u)dM(u)$$

is a $(\mathcal{G}_t)_{t\ge0}$-martingale. This concludes the proof. $\square$

Remark 5.41

If the condition that $X(\tau)$ should be integrable is relaxed in Theorem 5.40, but the remaining conditions for X hold true, then $\int_0^t X(u)dM(u)$ is a local martingale; see Definition A.21. The localising stopping times can be taken as

$$\tau_n = \inf\{t \geq 0 : |X(t)| \leq n\}.$$

6

Security pricing with hazard process

6.1   Bond price dynamics

Assumption 5.2, according to which the discounted defaultable bond price process is a martingale under the risk-neutral probability Q, allows us to write down a convenient formula for the bond price:

$$D(t, T) = e^{-r(T-t)} \mathbb{E}(D(T, T)|\mathcal{G}_t).$$

Recall that $\mathbb{E}$ denotes expectation under Q. The bond pays $D(T, T) = \mathbf{1}_{\{T < \tau\}}$ at maturity, hence Theorem 5.35 gives

$$
\begin{aligned}
D(t, T) &= e^{-r(T-t)} \mathbb{E}(\mathbf{1}_{\{T<\tau\}}|\mathcal{G}_t) \\
&= e^{-r(T-t)} \mathbf{1}_{\{t<\tau\}} \mathbb{E}(e^{-(\Gamma(T)-\Gamma(t))}|\mathcal{F}_t) \\
&= e^{-r(T-t)} \mathbf{1}_{\{t<\tau\}} e^{\Gamma(t)} \mathbb{E}(e^{-\Gamma(T)}|\mathcal{F}_t).
\end{aligned}
\tag{6.1}
$$

Remark 6.1

We can write the formula for the bond price in terms of the hazard rate γ

as

$$D(t, T) = e^{-r(T-t)} \mathbf{1}_{\{t<\tau\}} \mathbb{E}\left(\exp\left(-\int_t^T \gamma(u)du \right) \middle| \mathcal{F}_t \right)$$

$$= \mathbf{1}_{\{t<\tau\}} \mathbb{E}\left(\exp\left(-\int_t^T (r + \gamma(u))du \right) \middle| \mathcal{F}_t \right).$$

This formula is similar to that for non-defaultable bonds, but with the interest rate r increased by the hazard rate γ. In other words, the hazard rate plays the role of credit spread. If we specify a model of the hazard rate, the right-hand side can be computed and compared with market data to enable calibration.

Definition 3.12 of pre-default value can be extended as follows.

Definition 6.2

Let X be a $(\mathcal{G}_t)_{t\geq0}$-adapted process. The corresponding **pre-default value** process $\hat{X}$ is an $(\mathcal{F}_t)_{t\geq0}$-adapted process such that, for all $t \geq 0$,

$$X(t)\mathbf{1}_{\{t<\tau\}} = \hat{X}(t)\mathbf{1}_{\{t<\tau\}}.$$

When $X(t)$ is integrable, by taking the conditional expectation on both sides of this equality, we get

$$\mathbb{E}(X(t)\mathbf{1}_{\{t<\tau\}}|\mathcal{F}_t) = \mathbb{E}(\hat{X}(t)\mathbf{1}_{\{t<\tau\}}|\mathcal{F}_t) = \mathbb{E}(\mathbf{1}_{\{t<\tau\}}|\mathcal{F}_t)\hat{X}(t) = e^{-\Gamma(t)}\hat{X}(t)$$

since $\hat{X}(t)$ is $\mathcal{F}_t$-measurable. It means that $\hat{X}(t)$ is uniquely defined and can be expressed as

$$\hat{X}(t) = e^{\Gamma(t)}\mathbb{E}(X(t)\mathbf{1}_{\{t<\tau\}}|\mathcal{F}_t).$$

As a result, for $X(t) = D(t, T)$ given by (6.1), we obtain

$$\hat{D}(t, T) = e^{\Gamma(t)}\mathbb{E}(D(t, T)\mathbf{1}_{\{t<\tau\}}|\mathcal{F}_t)$$

$$= e^{\Gamma(t)}e^{-r(T-t)}e^{\Gamma(t)}\mathbb{E}(e^{-\Gamma(T)}|\mathcal{F}_t)\mathbb{E}(\mathbf{1}_{\{t<\tau\}}|\mathcal{F}_t)$$

$$= e^{-r(T-t)}e^{\Gamma(t)}\mathbb{E}(e^{-\Gamma(T)}|\mathcal{F}_t).$$

We would like to derive a stochastic integral equation for $D(t, T)$. But first we focus on $\hat{D}(t, T)$, which is an Itô process, hence technically easier to handle. In fact, it is the product of two Itô processes, namely $e^{-r(T-t)}e^{\Gamma(t)}$ and $\mathbb{E}(e^{-\Gamma(T)}|\mathcal{F}_t) = \mathbb{E}(G(T)|\mathcal{F}_t)$. The former can be written as

$$e^{-r(T-t)}e^{\Gamma(t)} = 1 + \int_0^t (r + \gamma(u))e^{-r(T-u)}e^{\Gamma(u)}du.$$

The latter is an $(\mathcal{F}_t)_{t\geq0}$-martingale under Q. By the classical martingale representation theorem (see [BSM]), there is a process $X_{G(T)}$ of class $\mathcal{M}^2$ such that

$$\mathbb{E}(G(T)|\mathcal{F}_t) = \mathbb{E}(G(T)) + \int_0^t X_{G(T)}(u)dW_Q(u),$$

where W_Q is the Wiener process under Q given by (5.2). Note that

$$e^{-rT}\mathbb{E}(G(T)) = e^{-rT}Q(T < \tau) = D(0, T).$$

By Itô's product rule (see [SCF]),

$$\hat{D}(t, T) = D(0, T) + \int_0^t (r + \gamma(u))\hat{D}(u, T)du$$

$$+ \int_0^t e^{-r(T-u)}e^{\Gamma(u)}X_{G(T)}(u)dW_Q(u).$$

By Corollary 5.9, W_Q is a Wiener process with respect to $(\mathcal{G}_t)_{t\geq 0}$ under Q, hence we can insert the $(\mathcal{G}_t)_{t\geq 0}$-stopping time $t \wedge \tau$ for the upper integration limit. Since $D(u, T) = \mathbf{1}_{\{u<\tau\}}\hat{D}(u, T)$, we get

$$\hat{D}(t \wedge \tau, T) = D(0, T) + \int_0^{t\wedge\tau} (r + \gamma(u))\hat{D}(u, T)du$$

$$+ \int_0^{t\wedge\tau} e^{-r(T-u)}e^{\Gamma(u)}X_{G(T)}(u)dW_Q(u)$$

$$= D(0, T) + \int_0^t (r + \gamma(u))D(u, T)du$$

$$+ \int_0^{t\wedge\tau} e^{-r(T-u)}e^{\Gamma(u)}X_{G(T)}(u)dW_Q(u). \qquad (6.2)$$

We are ready to prove the following result, which extends Theorem 3.21 (see also Remark 3.22).

Theorem 6.3
The defaultable bond prices satisfy the equation

$$D(t, T) - D(0, T) = \int_0^t rD(u, T)du - \int_0^t D(u_-, T)dM(u)$$

$$+ \int_0^{t\wedge\tau} e^{-r(T-u)}e^{\Gamma(u)}X_{G(T)}(u)dW_Q(u). \qquad (6.3)$$

Proof The expression $\hat{D}(t \wedge \tau, T)$ on the left-hand side of (6.2) can be written as

$$\hat{D}(t \wedge \tau, T) = \mathbf{1}_{\{\tau\leq t\}}\hat{D}(\tau, T) + \mathbf{1}_{\{t<\tau\}}\hat{D}(t, T)$$

$$= \mathbf{1}_{\{\tau\leq t\}}\hat{D}(\tau, T) + D(t, T).$$

Formula (5.11) applied to the left limit $D(t_-, T) = \lim_{u \nearrow t} D(u, T)$ gives

$$\int_0^t D(u_-, T)dM(u) = \mathbf{1}_{\{\tau \le t\}} D(\tau_-, T) - \int_0^{t \wedge \tau} D(u_-, T)\gamma(u)du$$

$$= \mathbf{1}_{\{\tau \le t\}} \hat{D}(\tau, T) - \int_0^t D(u, T)\gamma(u)du.$$

Combining these formulae with (6.2) proves the theorem. □

Proposition 3.23 can be extended in a similar manner, and this is left as an exercise.

Exercise 6.1 Show that the discounted defaultable bond prices $\tilde{D}(t, T) = e^{-rt}D(t, T)$ satisfy the equation

$$\tilde{D}(t, T) - \tilde{D}(0, T) = -\int_0^t \tilde{D}(u_-, T)dM(u)$$

$$+ \int_0^{t \wedge \tau} e^{-rT} e^{\Gamma(u)} X_{G(T)}(u)dW_Q(u).$$

Remark 6.4
The equation for $\tilde{D}(t, T)$ in Exercise 6.1 is consistent with the fact that $\tilde{D}(t, T)$ is a $(\mathcal{G}_t)_{t \ge 0}$-martingale under Q (see Assumption 5.2), as both integrals on the right are $(\mathcal{G}_t)_{t \ge 0}$-martingales.

To conclude this section, we show that the process $X_{G(T)}$ featuring in the stochastic integral equation for $D(t, T)$ can be expressed in terms of the strategy replicating $G(T)$ in the Black–Scholes model.

Proposition 6.5
Let (α_B, α_S) be an admissible self-financing strategy that replicates the contingent claim $G(T) = \mathbb{E}(\mathbf{1}_{\{T < \tau\}} | \mathcal{F}_T)$ in the Black–Scholes model consisting of the assets B and S; see [BSM]. Then

$$X_{G(T)}(t) = \sigma e^{r(T-t)} \alpha_S(t) S(t),$$

and the stochastic integral equation (6.3) for $D(t, T)$ can be written as

$$D(t, T) - D(0, T) = \int_0^t rD(u, T)du - \int_0^t D(u_-, T)dM(u)$$

$$+ \int_0^{t \wedge \tau} e^{\Gamma(u)} \alpha_S(u) \sigma S(u)dW_Q(u).$$

Proof Let

$$V_\alpha(t) = \alpha_B(t)B(t,T) + \alpha_S(t)S(t)$$

denote the value of the strategy (α_B, α_S) at time t. Since the strategy is self-financing in the Black–Scholes model, the discounted value $\tilde{V}_\alpha(t) = e^{-rt}V_\alpha(t)$ can be expressed as

$$\tilde{V}_\alpha(t) = V_\alpha(0) + \int_0^t \alpha_S(u)d\tilde{S}(u)$$

$$= V_\alpha(0) + \int_0^t \alpha_S(u)\sigma e^{-ru}S(u)dW_Q(u).$$

By replication, $V_\alpha(T) = G(T)$. The discounted value $\tilde{V}_\alpha(t)$ is an $(\mathcal{F}_t)_{t\geq 0}$-martingale under Q. It follows that

$$\tilde{V}_\alpha(t) = \mathbb{E}(e^{-rT}V_\alpha(T)|\mathcal{F}_t)$$

$$= e^{-rT}\mathbb{E}(G(T)|\mathcal{F}_t)$$

$$= e^{-rT}\mathbb{E}(G(T)) + e^{-rT}\int_0^t X_{G(T)}(u)dW_Q(u)$$

for each $t \in [0,T]$. In particular, $\tilde{V}_\alpha(0) = e^{-rT}\mathbb{E}(G(T))$, so

$$\int_0^t \alpha_S(u)\sigma e^{-ru}S(u)dW_Q(u) = e^{-rT}\int_0^t X_{G(T)}(u)dW_Q(u)$$

for each $t \in [0,T]$. As a result,

$$\alpha_S(t)\sigma e^{-rt}S(t) = e^{-rT}X_{G(T)}(t),$$

which implies the claimed formula for $X_{G(T)}(t)$. Finally, by inserting this formula into (6.3), we obtain the equation for $D(t,T)$. □

6.2 Trading strategies

In this section we extend the notion of self-financing strategies to the market with a default-free bond B and risky asset S, which follow the Black–Scholes model, and a defaultable bond D. We also discuss replicating strategies and touch upon the notion of arbitrage in this setting.

Definition 6.6

A trading **strategy** in the market model consisting of the assets B, S, D is a triple $\varphi = (\varphi_B, \varphi_S, \varphi_D)$ of left-continuous $(\mathcal{G}_t)_{t \geq 0}$-adapted processes. The **value** of the strategy is defined as

$$V_\varphi(t) = \varphi_B(t)B(t, T) + \varphi_S(t)S(t) + \varphi_D(t)D(t, T),$$

and the discounted value as

$$\tilde{V}_\varphi(t) = e^{-rt}V_\varphi(t)$$

for each $t \in [0, T]$.

The self-financing property of a strategy can be captured as

$$dV_\varphi(t) = \varphi_B(t)dB(t, T) + \varphi_S(t)dS(t) + \varphi_D(t)dD(t, T),$$

an equality in differential form, extending an analogous formula in the Black–Scholes model, where the last term on the right is absent; see [BSM]. To make the equality rigorous, we need to write it in integral form. The first two terms on the right become

$$\int_0^t \varphi_B(u)dB(u, T) = \int_0^t \varphi_B(t)re^{-r(T-u)}du,$$

$$\int_0^t \varphi_S(u)dS(u) = \int_0^t \varphi_S(u)rS(u)du + \int_0^t \varphi_S(u)\sigma S(u)dW_Q(u),$$

just like in the Black–Scholes model. In the last term we can employ the informal formula

$$dD(t, T) = rD(t, T)dt - D(t_-, T)dM(t) + \mathbf{1}_{\{t < \tau\}}e^{-r(T-t)}e^{\Gamma(t)}X_{G(T)}(t)dW_Q(t)$$

implicated by Theorem 6.3, but need to treat it carefully when converting into a rigorous integral form because both $D(t, T)$ and $M(t)$ have a jump at time τ. To this end we consider two cases covering the pre-default and post-default regions $\{t < \tau\}$ and $\{\tau \leq t\}$.

Definition 6.7
A trading strategy $(\varphi_B, \varphi_S, \varphi_D)$ is called **self-financing** if for all $t \in [0, T]$

(i) in the pre-default region $\{t < \tau\}$

$$V_\varphi(t) = V_\varphi(0) + \int_0^t \varphi_B(u) r B(u, T) du$$

$$+ \int_0^t \varphi_S(u) r S(u) du + \int_0^t \varphi_S(u) \sigma S(u) dW_Q(u)$$

$$+ \int_0^t \varphi_D(u) r D(u, T) du + \int_0^t \varphi_D(u) \gamma(u) D(u, T) du$$

$$+ \int_0^t \varphi_D(u) e^{-r(T-u)} e^{\Gamma(u)} X_{G(T)}(u) dW_Q(u);$$

(ii) in the post-default region $\{\tau \leq t\}$

$$V_\varphi(t) = V_\varphi(\tau) + \int_\tau^t \varphi_B(u) r B(u, T) du$$

$$+ \int_\tau^t \varphi_S(u) r S(u) du + \int_\tau^t \varphi_S(u) \sigma S(u) dW_Q(u).$$

Remark 6.8

In Definition 6.7 the stochastic integrals with respect to W_Q are well defined and can be restricted to the events $\{t < \tau\}, \{\tau \leq t\} \in \mathcal{G}_t$, and the $(\mathcal{G}_t)_{t \geq 0}$-stopping time τ can be used as an integration limit because W_Q is a Wiener process with respect to $(\mathcal{G}_t)_{t \geq 0}$, as shown in Corollary 5.9.

Proposition 6.9

A necessary and sufficient condition for a strategy $(\varphi_B, \varphi_S, \varphi_D)$ to be self-financing is that the discounted value process can be expressed as

$$\tilde{V}_\varphi(t) = V_\varphi(0) + \int_0^t e^{-ru} \varphi_S(u) \sigma S(u) dW_Q(u)$$

$$- \int_0^t e^{-ru} \varphi_D(u) D(u_-, T) dM(u)$$

$$+ e^{-rT} \int_0^{t \wedge \tau} \varphi_D(u) e^{\Gamma(u)} X_{G(T)}(u) dW_Q(u) \qquad (6.4)$$

for all $t \in [0, T]$.

Proof First, we prove that the expression for $\tilde{V}_\varphi(t)$ is a necessary condition for self-financing. On $\{t < \tau\}$, Itô's product rule applied to $e^{-rt} V_\varphi(t)$

gives

$$e^{-rt}V_\varphi(t) = V_\varphi(0) - \int_0^t e^{-ru} rV_\varphi(u)du + \int_0^t e^{-ru}\varphi_B(u)rB(u,T)du$$

$$+ \int_0^t e^{-ru}\varphi_S(u)rS(t)dt + \int_0^t e^{-ru}\varphi_S(u)\sigma S(u)dW_Q(u)$$

$$+ \int_0^t e^{-ru}\varphi_D(u)rD(u,T)du + \int_0^t e^{-ru}\varphi_D(u)\gamma(u)D(u,T)du$$

$$+ \int_0^t e^{-ru}\varphi_D(u)e^{-r(T-u)}e^{\Gamma(u)}X_{G(T)}(u)dW_Q(u)$$

$$= V_\varphi(0) + \int_0^t e^{-ru}\varphi_S(u)\sigma S(u)dW_Q(u)$$

$$+ \int_0^t e^{-ru}\varphi_D(u)\gamma(u)D(u,T)du$$

$$+ \int_0^t e^{-ru}\varphi_D(u)e^{-r(T-u)}e^{\Gamma(u)}X_{G(T)}(u)dW_Q(u). \qquad (6.5)$$

Similarly, in the post-default region $\{\tau \le t\}$, Itô's product rule gives

$$e^{-rt}V_\varphi(t) = e^{-r\tau}V_\varphi(\tau) - r\int_\tau^r e^{-ru}V_\varphi(u)du + \int_\tau^t e^{-ru}\varphi_B(u)rB(u,T)du$$

$$+ \int_\tau^t e^{-ru}\varphi_S(u)rS(u)du + \int_\tau^t e^{-ru}\varphi_S(u)\sigma S(u)dW_Q(u)$$

$$= e^{-r\tau}V_\varphi(\tau) + \int_\tau^t e^{-ru}\varphi_S(u)\sigma S(u)dW_Q(u). \qquad (6.6)$$

By taking the left limit of $V_\varphi(t) = \varphi_B(t)B(t,T) + \varphi_S(t)S(t) + \varphi_D(t)D(t,T)$ as $t \nearrow \tau$, we get

$$V_\varphi(\tau_-) = \varphi_B(\tau)B(\tau,T) + \varphi_S(\tau)S(\tau) + \varphi_D(\tau)D(\tau_-,T)$$

$$= V_\varphi(\tau) + \varphi_D(\tau)D(\tau_-,T).$$

On the other hand, the left limit as $t \nearrow \tau$ in (6.5) gives

$$e^{-r\tau}V_\varphi(\tau_-) = V_\varphi(0) + \int_0^\tau e^{-ru}\varphi_S(u)\sigma S(u)dW_Q(u)$$

$$+ \int_0^\tau e^{-ru}\varphi_D(u)\gamma(u)D(u,T)du$$

$$+ \int_0^\tau e^{-ru}\varphi_D(u)e^{-r(T-u)}e^{\Gamma(u)}X_{G(T)}(u)dW_Q(u).$$

It follows that

$$e^{-r\tau}V_\varphi(\tau) = e^{-r\tau}V_\varphi(\tau_-) - e^{-r\tau}\varphi_D(\tau)D(\tau_-, T)$$

$$= V_\varphi(0) + \int_0^\tau e^{-ru}\varphi_S(u)\sigma S(u)dW_Q(u)$$

$$+ \int_0^\tau e^{-ru}\varphi_D(u)\gamma(u)D(u, T)du - e^{-r\tau}\varphi_D(\tau)D(\tau_-, T)$$

$$+ \int_0^\tau e^{-ru}\varphi_D(u)e^{-r(T-u)}e^{\Gamma(u)}X_{G(T)}(u)dW_Q(u).$$

By inserting this into (6.6), we obtain

$$e^{-rt}V_\varphi(t) = e^{-r\tau}V_\varphi(\tau) + \int_\tau^t e^{-ru}\varphi_S(u)\sigma S(u)dW_Q(u)$$

$$= V_\varphi(0) + \int_0^t e^{-ru}\varphi_S(u)\sigma S(u)dW_Q(u)$$

$$+ \int_0^\tau e^{-ru}\varphi_D(u)\gamma(u)D(u, T)du - e^{-r\tau}\varphi_D(\tau)D(\tau_-, T)$$

$$+ \int_0^\tau e^{-ru}\varphi_D(u)e^{-r(T-u)}e^{\Gamma(u)}X_{G(T)}(u)dW_Q(u)$$

in $\{\tau \le t\}$. Combining the last expression with (6.5), we finally get

$$e^{-rt}V_\varphi(t) = V_\varphi(0) + \int_0^t e^{-ru}\varphi_S(u)\sigma S(u)dW_Q(u)$$

$$+ \int_0^{t\wedge\tau} e^{-ru}\varphi_D(u)\gamma(u)D(u_-, T)du - 1_{\{\tau\le t\}}e^{-r\tau}\varphi_D(\tau)D(\tau_-, T)$$

$$+ \int_0^{t\wedge\tau} e^{-ru}\varphi_D(u)e^{-r(T-u)}e^{\Gamma(u)}X_{G(T)}(u)dW_Q(u)$$

$$= V_\varphi(0) + \int_0^t e^{-ru}\varphi_S(u)\sigma S(u)dW_Q(u)$$

$$- \int_0^t e^{-ru}\varphi_D(u)D(u_-, T)dM(u)$$

$$+ e^{-rT}\int_0^{t\wedge\tau} \varphi_D(u)e^{\Gamma(u)}X_{G(T)}(u)dW_Q(u)$$

for all $t \in [0, T]$. This completes the proof that the formula for $\tilde{V}_\varphi(t)$ is a necessary condition for self-financing. By reversing the above argument, we can show that it is also a necessary one.   □

Exercise 6.2 Let $(\varphi_B, \varphi_S, \varphi_D)$ be a self-financing strategy. Show that for all $t \in [0, T]$

$$V_\varphi(t) = V_\varphi(0) + \int_0^t rV_\varphi(u)du + \int_0^t \varphi_S(u)\sigma S(u)dW_Q(u)$$

$$- \int_0^t \varphi_D(u)D(u-, T)dM(u)$$

$$+ \int_0^{t \wedge \tau} \varphi_D(u)e^{-r(T-u)}e^{\Gamma(u)}X_{G(T)}(u)dW_Q(u).$$

In the expression for $\tilde{V}_\varphi(t)$ in Proposition 6.9, the integrals with respect to W_Q and M are $(\mathcal{G}_t)_{t\geq 0}$-martingales if the integrands are regular enough, but this does not have to be the case for all self-financing strategies. Just like in Black–Scholes theory (see [BSM]), to exclude pathological situations such as doubling strategies or suicide strategies we need an additional condition.

Definition 6.10

We say that a self-financing strategy $(\varphi_B, \varphi_S, \varphi_D)$ is **admissible** if the discounted value process $\tilde{V}_\varphi$ is a $(\mathcal{G}_t)_{t\geq 0}$-martingale under Q.

The following notions and results are reformulations of the corresponding items in the Black–Scholes model.

Definition 6.11

An **arbitrage strategy** in the market model consisting of the assets B, S, D is a self-financing strategy $(\varphi_B, \varphi_S, \varphi_D)$ such that $V_\varphi(0) = 0$, $V_\varphi(T) \geq 0$, and $V_\varphi(T) > 0$ with positive probability for some $T > 0$.

Theorem 6.12

In the market model consisting of the assets B, S, D there is no admissible arbitrage strategy.

Proof Suppose that $V_\varphi(T) \geq 0$ and $V_\varphi(0) = 0$ for some admissible self-financing strategy $(\varphi_B, \varphi_S, \varphi_D)$. The discounted value process $\tilde{V}_\varphi$ is a $(\mathcal{G}_t)_{t\geq 0}$-martingale under Q, hence $e^{-rT}\mathbb{E}(V_\varphi(T)) = V_\varphi(0) = 0$. Because $V_\varphi(T) \geq 0$, this implies that $Q(V_\varphi(T) > 0) = 0$. It follows that an arbitrage strategy cannot be found. $\square$

Exercise 6.3   Show that if a self-financing strategy φ in the model consisting of the assets B, S, D has non-negative values, $V_\varphi(t) \geq 0$ for all $t \geq 0$, then φ cannot be an arbitrage strategy.

Next we consider a contingent claim with exercise time $T > 0$ whose payoff is a $\mathcal{G}_T$-measurable random variable H. We denote the price process of the contingent claim by $H(t)$, hence $H(T) = H$.

Definition 6.13
A strategy $\varphi = (\varphi_B, \varphi_S, \varphi_D)$ is said to **replicate** the contingent claim H whenever $V_\varphi(T) = H(T)$.

In the extended market consisting of the assets B, S, D and H we shall only use strategies with static positions in the contingent claim H. In fact, to specify the self-financing condition for strategies with more general non-static positions in the contingent claim we would need to know more about the dynamics of $H(t)$ so as to be able to write down a stochastic integral with respect to $H(t)$.

Definition 6.14
By an **arbitrage opportunity** in the extended market model consisting of the assets B, S, D and contingent claim H we understand a self-financing strategy $\psi = (\psi_B, \psi_S, \psi_D)$ in the B, S, D market and a static position $z = 1$ or -1 in the contingent claim H such that $V_\psi(0) + zH(0) = 0$, $V_\psi(T) + zH(T) \geq 0$, and $V_\psi(T) + zH(T) > 0$ with positive probability.

Theorem 6.15
Let $\varphi = (\varphi_B, \varphi_S, \varphi_D)$ be a self-financing strategy in the B, S, D market that replicates a contingent claim with $\mathcal{G}_T$-measurable payoff H and exercise time T. If there are no arbitrage opportunities in the extended market, then at any time $t \in [0, T]$ the price of the contingent claim must be equal to the value of the replicating strategy, that is,

$$H(t) = V_\varphi(t).$$

Proof   It is enough to verify this equality for $t = 0$, given that we can choose, without loss of generality, the initial time 0 to be any time instant before the exercise date of the contingent claim.

Suppose that $H(0) > V_\varphi(0)$. In this case enter a long position in the replicating strategy, a short position in the contingent claim, and invest the difference $H(0) - V_\varphi(0) > 0$ in the non-defaultable bond. In other words, follow the self-financing strategy in the B, S, D market given by

$$(\psi_B(t), \psi_S(t), \psi_D(t)) = (\varphi_B(t) + B(0, T)^{-1}(H(0) - V_\varphi(0)), \varphi_S(t), \varphi_D(t))$$

for each $t \in [0, T]$, and take the position $z = -1$ in the contingent claim. Then

$$V_\psi(0) + zH(0) = V_\varphi(0) + (H(0) - V_\varphi(0)) - H(0) = 0$$

and

$$\begin{aligned} V_\psi(T) + zH(T) &= V_\varphi(T) + B(0, T)^{-1}(H(0) - V_\varphi(0)) - H(T) \\ &= B(0, T)^{-1}(H(0) - V_\varphi(0)) > 0 \end{aligned}$$

since the value $V_\varphi(T)$ of the replicating strategy is equal to $H(T)$. This is an arbitrage opportunity in the extended market model. When $H(0) > V_\varphi(0)$, the opposite positions produce an arbitrage opportunity. Hence the equality $H(0) = V_\varphi(0)$ must hold if there are no arbitrage opportunities in the extended market. $\qquad\qquad\square$

6.3 Zero-recovery securities

Let X be a non-negative $\mathcal{F}_T$-measurable square-integrable random variable representing the payoff at time $T > 0$ of a contingent claim in the case when there is no default before or at time T. If a default occurs before or at time T, the payoff will be 0 (zero recovery). The payoff of such a contingent claim can be represented as $X\mathbf{1}_{\{T < \tau\}}$, a $\mathcal{G}_T$-measurable random variable. We are going to derive formulae for a strategy replicating this payoff.

To replicate the payoff $X\mathbf{1}_{\{T < \tau\}}$ it will be sufficient to find a self-financing strategy $\varphi = (\varphi_B, \varphi_S, \varphi_D)$ whose values satisfy

$$V_\varphi(t) = e^{-r(T-t)}\mathbb{E}(X\mathbf{1}_{\{T < \tau\}}|\mathcal{G}_t)$$

for all $t \in [0, T]$. Indeed, for $t = T$, this gives the replication condition $V_\varphi(T) = X\mathbf{1}_{\{T < \tau\}}$. The strategy will be admissible because $\tilde{V}_\varphi(t) = e^{-rT}\mathbb{E}(X\mathbf{1}_{\{T < \tau\}}|\mathcal{G}_t)$ is a $(\mathcal{G}_t)_{t \geq 0}$-martingale.

By Theorem 5.35, for any $t \in [0, T]$

$$\mathbb{E}(X\mathbf{1}_{\{T < \tau\}}|\mathcal{G}_t) = \mathbf{1}_{\{t < \tau\}}e^{\Gamma(t)}\mathbb{E}(Xe^{-\Gamma(T)}|\mathcal{F}_t).$$

Moreover, according to (6.1),

$$D(t, T) = e^{-r(T-t)}\mathbf{1}_{\{t < \tau\}}e^{\Gamma(t)}\mathbb{E}(e^{-\Gamma(T)}|\mathcal{F}_t).$$

Observe that $e^{-\Gamma(T)}$ is an $\mathcal{F}_T$-measurable random variable, which can be regarded as the payoff of a derivative security in the default-free Black–Scholes market with assets B and S. Let (α_B, α_S) be an admissible self-financing strategy replicating $e^{-\Gamma(T)}$ in this market (see [BSM]). The values

$$V_\alpha(t) = \alpha_B(t)B(t, T) + \alpha_S(t)S(t)$$

of this strategy satisfy

$$V_\alpha(t) = e^{-r(T-t)}\mathbb{E}(e^{-\Gamma(T)}|\mathcal{F}_t).$$

Similarly, $Xe^{-\Gamma(T)}$ is an $\mathcal{F}_T$-measurable random variable and can also be regarded as the payoff of a derivative security in the default-free Black–Scholes market. Let (β_B, β_S) be an admissible self-financing strategy replicating $Xe^{-\Gamma(T)}$ in this market, whose values

$$V_\beta(t) = \beta_B(t)B(t, T) + \beta_S(t)S(t)$$

satisfy

$$V_\beta(t) = e^{-r(T-t)}\mathbb{E}(Xe^{-\Gamma(T)}|\mathcal{F}_t).$$

Hence, it will be enough to construct a self-financing strategy $(\varphi_B, \varphi_S, \varphi_D)$ such that

$$
\begin{aligned}
V_\varphi(t) &= e^{-r(T-t)}\mathbb{E}(X\mathbf{1}_{\{T<\tau\}}|\mathcal{G}_t)\\
&= e^{-r(T-t)}\mathbf{1}_{\{t<\tau\}}e^{\Gamma(t)}\mathbb{E}(Xe^{-\Gamma(T)}|\mathcal{F}_t)\\
&= D(t, T)\frac{\mathbb{E}(Xe^{-\Gamma(T)}|\mathcal{F}_t)}{\mathbb{E}(e^{-\Gamma(T)}|\mathcal{F}_t)}\\
&= D(t, T)\frac{V_\beta(t)}{V_\alpha(t)}.
\end{aligned}
$$

This suggests that we should take

$$\varphi_D(t) = \frac{V_\beta(t)}{V_\alpha(t)} \tag{6.7}$$

and $\varphi_B(t), \varphi_S(t)$ such that

$$\varphi_B(t)B(t, T) + \varphi_S(t)S(t) = 0.$$

Because $D(t, T)$ becomes zero at the default time τ, which can take any positive value, it is natural to require that the last equality should be satisfied for all $t \in [0, T]$ so as to have zero terminal payoff on $\{\tau \leq T\}$. We also need to ensure that the strategy $(\varphi_B, \varphi_S, \varphi_D)$ is self-financing. This will give an expression for $\varphi_S(t)$, and finally for $\varphi_B(t)$, which can be written as

$$\varphi_B(t) = -\frac{\varphi_S(t)S(t)}{B(t, T)}. \tag{6.8}$$

Observe that

$$
\begin{aligned}
V_\varphi(t) &= e^{-r(T-t)}\mathbb{E}(X\mathbf{1}_{\{T<\tau\}}|\mathcal{G}_t) \\
&= e^{-r(T-t)}\mathbf{1}_{\{t<\tau\}}e^{\Gamma(t)}\mathbb{E}(Xe^{-\Gamma(T)}|\mathcal{F}_t) \\
&= \mathbf{1}_{\{t<\tau\}}e^{\Gamma(t)}V_\beta(t).
\end{aligned}
$$

In the pre-default region $\{t < \tau\}$ this becomes

$$
V_\varphi(t) = e^{\Gamma(t)}V_\beta(t).
$$

The self-financing condition for the strategy (β_B, β_S) in the Black–Scholes model gives

$$
\begin{aligned}
V_\beta(t) &= V_\beta(0) + \int_0^t \beta_B(u)dB(u,T) + \int_0^t \beta_S(u)dS(u) \\
&= V_\beta(0) + \int_0^t \beta_B(u)rB(u,T)du \\
&\quad + \int_0^t \beta_S(u)rS(u)du + \int_0^t \beta_S(u)\sigma S(u)dW_Q(u) \\
&= V_\beta(0) + \int_0^t rV_\beta(u)du + \int_0^t \beta_S(u)\sigma S(u)dW_Q(u).
\end{aligned}
$$

By the Itô product rule,

$$
\begin{aligned}
V_\varphi(t) &= e^{\Gamma(t)}V_\beta(t) \\
&= V_\beta(0) + \int_0^t (r+\gamma(u))e^{\Gamma(u)}V_\beta(u)du + \int_0^t e^{\Gamma(u)}\beta_S(u)\sigma S(u)dW_Q(u)
\end{aligned}
$$
$$(6.9)$$

in the pre-default region $\{t < \tau\}$. On the other hand, according to Definition 6.7 and Proposition 6.5, the self-financing condition for the strategy $(\varphi_B, \varphi_S, \varphi_D)$ can be written as

$$
\begin{aligned}
V_\varphi(t) &= V_\varphi(0) + \int_0^t rV_\varphi(u)du + \int_0^t \varphi_D(u)\gamma(u)D(u,T)du \\
&\quad + \int_0^t \varphi_S(u)\sigma S(u)dW_Q(u) + \int_0^t \varphi_D(u)e^{\Gamma(u)}\sigma\alpha_S(u)S(u)dW_Q(u) \\
&= V_\varphi(0) + \int_0^t \left(re^{\Gamma(u)}V_\beta(u) + \varphi_D(u)\gamma(u)D(u,T)\right)du \\
&\quad + \int_0^t \left(\varphi_S(u)\sigma S(u) + \varphi_D(u)e^{\Gamma(u)}\sigma\alpha_S(u)S(u)\right)dW_Q(u) \quad (6.10)
\end{aligned}
$$

in the pre-default region $\{t < \tau\}$. By comparing the above expressions for

$V_\varphi(t)$, we can see that the integrals with respect to u are equal to one another when $\varphi_D(t)$ is given by (6.7); see Exercise 6.4. To make the stochastic integrals equal to one another we can take $\varphi_S(t)$ such that

$$e^{\Gamma(t)}\beta_S(t)\sigma S(t) = \varphi_S(t)\sigma S(t) + \varphi_D(t)e^{\Gamma(t)}\sigma\alpha_S(t)S(t).$$

This gives

$$\varphi_S(t) = e^{\Gamma(t)}(\beta_S(t) - \varphi_D(t)\alpha_S(t))$$

in $\{t < \tau\}$.

In the post-default region $\{\tau \le t\}$

$$V_\varphi(t) = \varphi_B(t)B(t,T) + \varphi_S(t)S(t) = 0.$$

In particular, $V_\varphi(\tau) = 0$. By Definition 6.7, the self-financing condition for $(\varphi_B, \varphi_S, \varphi_D)$ in the post-default region $\{\tau \le t\}$ can be written as

$$V_\varphi(t) = V_\varphi(\tau) + \int_\tau^t rV_\varphi(u)du + \int_\tau^t \varphi_S(u)\sigma S(u)dW_Q(u)$$

$$= \int_\tau^t rV_\varphi(u)du + \int_\tau^t \varphi_S(u)\sigma S(u)dW_Q(u).$$

This will indeed be equal to zero if we put

$$\varphi_S(t) = 0$$

on $\{\tau \le t\}$.

To summarise, for any $t \in [0,T]$ we take

$$\varphi_S(t) = \mathbf{1}_{\{t<\tau\}}e^{\Gamma(t)}(\beta_S(t) - \varphi_D(t)\alpha_S(t)),$$

with $\varphi_D(t)$ and $\varphi_B(t)$ given by (6.7) and (6.8). This ensures that $\varphi_B(t)$, $\varphi_S(t)$, $\varphi_D(t)$ are $\mathcal{G}_t$-measurable random variables for each t and satisfy the self-financing condition in Definition 6.7, and that $\tilde{V}_\varphi$ is a non-negative $(\mathcal{G}_t)_{t \ge 0}$-martingale; hence the admissibility condition in Definition 6.10 holds.

However, left-continuity of the paths of φ fails at time τ because of the indicator function $\mathbf{1}_{\{t<\tau\}}$ in the expression for $\varphi_S(t)$. This expression needs to be modified to ensure left-continuity without destroying the remaining properties for a replicating strategy. It can be done by replacing $\varphi_S(t)$ by the pathwise left limit $\varphi_S(t_-)$ for each $t > 0$. Because $\varphi_S(t - \frac{1}{n})$ is a $\mathcal{G}_{t-\frac{1}{n}}$-measurable random variable, it is $\mathcal{G}_t$-measurable for any $n = 1, 2, \ldots$, so the left limit $\varphi_S(t_-) = \lim_{n\to\infty} \varphi_S(t - \frac{1}{n})$ is also $\mathcal{G}_t$-measurable. The remaining random variables in the expression for $\varphi_S(t)$ have continuous paths, and hence are not affected by the limit. In particular, the stock positions $\alpha_S(t)$ and $\beta_S(t)$ in the Black–Scholes strategies replicating the payoffs $e^{-\Gamma(T)}$ and

$Xe^{-\Gamma(T)}$, respectively, have continuous paths since they can be expressed as partial derivatives with respect to the stock price of the corresponding solutions to the Black–Scholes partial differential equation (delta hedging; see [BSM]). After this modification, we get

$$\varphi_S(t) = \mathbf{1}_{\{t \le \tau\}} e^{\Gamma(t)} \left(\beta_S(t) - \varphi_D(t) \alpha_S(t) \right) \tag{6.11}$$

with $\varphi_D(t)$ and $\varphi_B(t)$ given by (6.7) and (6.8), respectively.

We have proved the following result.

Theorem 6.16
Formulae (6.7), (6.8), and (6.11) define an admissible self-financing strategy $(\varphi_B, \varphi_S, \varphi_D)$ that replicates the contingent claim $X\mathbf{1}_{\{T < \tau\}}$.

Exercise 6.4 Show that the integrals with respect to u in (6.9) and (6.10) are equal to one another when $\varphi_D(t)$ is given by (6.7).

Example 6.17
Take $H = (S(T) - K)^+ \mathbf{1}_{\{T < \tau\}}$, the payoff of a so-called **vulnerable call option**. The company issuing the option is subject to credit risk. The call payoff $(S(T) - K)^+$ is available only if the company is still operating at time T. The initial price of this derivative security is the expectation of the discounted payoff with respect to the risk-neutral probability,

$$\begin{aligned} H(0) &= e^{-rT} \mathbb{E}((S(T) - K)^+ \mathbf{1}_{\{T < \tau\}}) \\ &= e^{-rT} \mathbb{E}((S(T) - K)^+ \mathbb{E}(\mathbf{1}_{\{T < \tau\}} | \mathcal{F}_T)) \\ &= e^{-rT} \mathbb{E}((S(T) - K)^+ e^{-\Gamma(T)}). \end{aligned}$$

Suppose that the hazard process is as in Example 5.22,

$$\Gamma(t) = \max_{u \in [0,t]} \left(-\ln \frac{S(u)}{S(0)} \right) = -\min_{u \in [0,t]} R(u),$$

where

$$R(t) = \ln \frac{S(t)}{S(0)} = \left(r - \frac{1}{2}\sigma^2 \right) t + \sigma W_Q(t)$$

is the logarithmic return on stock in the Black–Scholes model. To compute the above expectation we need the joint distribution of $R(T)$

and $\min_{u\in[0,T]} R(u)$. This we can get from Proposition A.15. Namely, the joint density of $R(T)$ and $\min_{u\in[0,T]} R(u)$ is

$$f(x,y) = \frac{1}{\sigma\sqrt{2\pi T}} \frac{2(x-2y)}{T\sigma^2} e^{\frac{2y(r-\frac{1}{2}\sigma^2)}{\sigma^2}} e^{-\frac{(2y-x+rT-\frac{1}{2}\sigma^2 T)^2}{2\sigma^2 T}}$$

when $y \le 0$ and $y \le x$, and $f(x,y) = 0$ otherwise, and the vulnerable call price can be expressed as

$$
\begin{aligned}
H(0) &= e^{-rT}\mathbb{E}((S(T)-K)^+ e^{-\Gamma(T)}) \\
&= e^{-rT}\mathbb{E}((S(0)e^{R(T)}-K)^+ e^{\min_{u\in[0,T]} R(t)}) \\
&= e^{-rT}\int_{\ln\frac{K}{S(0)}}^{\infty}\int_{-\infty}^{0\wedge x}(S(0)e^x - K)e^y f(x,y)dydx.
\end{aligned}
$$

Exercise 6.5 Compute the price of a vulnerable call option with strike price $K = 100$ and exercise time $T = 1$ when $r = 5\%$, $S(0) = 100$, $\sigma = 30\%$, and the hazard process $\Gamma(t)$ is as in Example 6.17. Compare with the price of a standard call option in the Black–Scholes model.

6.4 Martingale representation theorem

Here we extend Theorem 4.1 to the market model consisting of the assets B, S, D. This will make it possible to replicate derivative securities in this model in a more systematic manner. In particular, equipped with the martingale representation theorem, we shall revisit zero-recovery securities, already studied in Section 6.3. Then, in Section 6.5, we are going to consider securities with positive recovery.

Theorem 6.18
For any $(\mathcal{F}_t)_{t\ge 0}$-adapted process $h(t), t \ge 0$, with left-continuous paths and such that $h(\tau)$ is square integrable with respect to Q, the $(\mathcal{G}_t)_{t\ge 0}$-martingale $H(t) = \mathbb{E}(h(\tau)|\mathcal{G}_t)$ can be represented as

$$H(t) = H(0) + \int_0^{t\wedge\tau} e^{\Gamma(s)}X(u)dW_Q(u) + \int_0^t (h(u) - J(u))dM(u),$$

where X is a process of class $\mathcal{M}^2$ such that the $(\mathcal{F}_t)_{t\geq 0}$-martingale

$$m(t) = \mathbb{E}(h(\tau)|\mathcal{F}_t)$$

can be represented as

$$m(t) = m(0) + \int_0^t X(u)dW_Q(u), \tag{6.12}$$

and where

$$J(t) = e^{\Gamma(t)}\mathbb{E}(\mathbf{1}_{\{t<\tau\}}h(\tau)|\mathcal{F}_t).$$

Proof It follows from the proof of Theorem 5.34 that

$$m(t) = \mathbb{E}(h(\tau)|\mathcal{F}_t)$$

$$= \mathbb{E}\left(\int_0^\infty h(u)f(u)du \,\Big|\, \mathcal{F}_t\right)$$

$$= \int_0^t h(u)f(u)du + \mathbb{E}\left(\int_t^\infty h(u)f(u)du \,\Big|\, \mathcal{F}_t\right)$$

$$= \int_0^t h(u)f(u)du + \mathbb{E}(\mathbf{1}_{\{t<\tau\}}h(\tau)|\mathcal{F}_t).$$

As a result,

$$J(t) = e^{\Gamma(t)}\mathbb{E}(\mathbf{1}_{\{t<\tau\}}h(\tau)|\mathcal{F}_t)$$

$$= e^{\Gamma(t)}\left(m(t) - \int_0^t h(u)f(u)du\right). \tag{6.13}$$

Hence, according to Theorem 5.34,

$$H(t) = \mathbb{E}(h(\tau)|\mathcal{G}_t)$$

$$= \mathbf{1}_{\{\tau\leq t\}}h(\tau) + \mathbf{1}_{\{t<\tau\}}e^{\Gamma(t)}\mathbb{E}\left(\int_t^\infty h(u)f(u)du \,\Big|\, \mathcal{F}_t\right)$$

$$= \mathbf{1}_{\{\tau\leq t\}}h(\tau) + \mathbf{1}_{\{t<\tau\}}e^{\Gamma(t)}\left(m(t) - \int_0^t h(u)f(u)du\right)$$

$$= \mathbf{1}_{\{\tau\leq t\}}h(\tau) + \mathbf{1}_{\{t<\tau\}}J(t).$$

Since $J(t)$ is given by (6.13) with $m(t)$ as in (6.12) and

$$e^{\Gamma(t)} = e^{\Gamma(0)} + \int_0^t \gamma(u)e^{\Gamma(u)}du,$$

it follows by the Itô product rule that

$$J(t) = J(0) + \int_0^t \gamma(u)e^{\Gamma(u)}\left(m(u) - \int_0^u h(s)f(s)ds\right)du$$

$$+ \int_0^t e^{\Gamma(u)}X(u)dW_Q(u) - \int_0^t e^{\Gamma(u)}h(u)f(u)du$$

$$= J(0) + \int_0^t e^{\Gamma(u)}X(u)dW_Q(u) - \int_0^t \gamma(u)(h(u) - J(u))du.$$

As a result,

$$H(t) = \mathbf{1}_{\{\tau \le t\}}h(\tau) + \mathbf{1}_{\{t < \tau\}}J(t)$$

$$= J(t \wedge \tau) + \mathbf{1}_{\{\tau \le t\}}(h(\tau) - J(\tau))$$

$$= J(0) + \int_0^{t \wedge \tau} e^{\Gamma(u)}X^h(u)dW_Q(u)$$

$$- \int_0^{t \wedge \tau} \gamma(u)(h(u) - J(u))du + \mathbf{1}_{\{\tau \le t\}}(h(\tau) - J(\tau))$$

$$= J(0) + \int_0^{t \wedge \tau} e^{\Gamma(u)}X^h(u)dW_Q(u) + \int_0^t (h(u) - J(u))dM(u)$$

for any $t \ge 0$. In particular, $H(0) = J(0)$, which completes the proof. $\square$

Remark 6.19

Because $h(\tau)$ is square integrable with respect to Q, we have

$$\mathbb{E}(m(t)^2) = \mathbb{E}(\mathbb{E}(h(\tau)|\mathcal{F}_t)^2) \le \mathbb{E}(\mathbb{E}(h(\tau)^2|\mathcal{F}_t)) = \mathbb{E}(h(\tau)^2) < \infty$$

by Jensen's inequality for conditional expectation; see [PF]. Hence $m(t)$ is square integrable with respect to Q, and it follows that the process X of class $\mathcal{M}^2$ in the representation (6.12) of the $(\mathcal{F}_t)_{t \ge 0}$-martingale $m(t) = \mathbb{E}(h(\tau)|\mathcal{F}_t)$ exists by the classical representation theorem used in the Black–Scholes model; see [BSM]. It can be expressed as

$$X(t) = e^{r(T-t)}\beta_S(t)\sigma S(t),$$

where (β_B, β_S) is the admissible self-financing strategy that replicates the derivative security with payoff $m(T) = \mathbb{E}(h(\tau)|\mathcal{F}_T)$ and exercise time T in the Black–Scholes model consisting of the assets B, S (the replicating strategy is the same for all $T > 0$). It follows that the $(\mathcal{G}_t)_{t \ge 0}$-martingale $H(t) = \mathbb{E}(h(\tau)|\mathcal{G}_t)$ can be represented as

$$H(t) = H(0) + \int_0^{t \wedge \tau} e^{\Gamma(u)}e^{r(T-u)}\beta_S(u)\sigma S(u)dW_Q(u) + \int_0^t (h(u) - J(u))dM(u).$$

Exercise 6.6 By Assumption 5.2, $e^{-rt}D(t, T)$ is a martingale with respect to the filtration $(\mathcal{G}_t)_{t \geq 0}$. Find the corresponding processes $h(t)$ and $J(t)$ in the representation of $e^{-rt}D(t, T)$ in Theorem 6.18.

Exercise 6.7 Find the processes $h(t)$ and $J(t)$ in the representation in Theorem 6.18 of the martingale $\mathbb{E}(X\mathbf{1}_{\{T < \tau\}}|\mathcal{G}_t)$, where $X\mathbf{1}_{\{T < \tau\}}$ is the payoff of a zero-recovery derivative security with X an $\mathcal{F}_T$-measurable random variable as considered in Section 6.3.

Next, let us consider the general case of a derivative security with payoff of the form $\mathbb{E}(h(\tau)|\mathcal{G}_T)$ and exercise time $T > 0$ for some $(\mathcal{F}_t)_{t \geq 0}$-adapted non-negative process h with left-continuous paths and such that $h(\tau)$ is square integrable. We are looking for an admissible self-financing strategy $(\varphi_B, \varphi_S, \varphi_D)$ that replicates this payoff. The value of this strategy should satisfy

$$V_\varphi(t) = e^{-r(T-t)}\mathbb{E}\left(h(\tau)|\mathcal{G}_t\right)$$

for all $t \in [0, T]$. According to Theorem 6.18 and Remark 6.19 as applied to the $(\mathcal{G}_t)_{t \geq 0}$-martingale $H(t) = \mathbb{E}(h(\tau)|\mathcal{G}_t)$, the discounted value $\tilde{V}_\varphi(t) = e^{-rt}V_\varphi(t) = e^{-rT}H(t)$ can be written as

$$\tilde{V}_\varphi(t) = V_\varphi(0) + \int_0^{t \wedge \tau} e^{\Gamma(u)}e^{-ru}\beta_S(u)\sigma S(u)dW_Q(u)$$
$$+ \int_0^t e^{-rT}(h(u) - J(u))dM(u), \tag{6.14}$$

where (β_B, β_S) is the admissible self-financing strategy that replicates the contingent claim with payoff $m(T) = \mathbb{E}(h(\tau)|\mathcal{F}_T)$ and exercise time T in the Black–Scholes model consisting of the assets B, S, and where $J(t)$ is given by (6.13). On the other hand, by Proposition 6.5 and Proposition 6.9 (i.e. the self-financing condition),

$$\tilde{V}_\varphi(t) = V_\varphi(0) + \int_0^t e^{-ru}\varphi_S(u)\sigma S(u)dW_Q(u)$$
$$- \int_0^t e^{-ru}\varphi_D(u)D(u_-, T)dM(u)$$
$$+ \int_0^{t \wedge \tau} e^{-ru}\varphi_D(u)e^{\Gamma(u)}\sigma\alpha_S(u)S(u)dW_Q(u), \tag{6.15}$$

where (α_B, α_S) is the admissible self-financing strategy that replicates the contingent claim with payoff $e^{-\Gamma(T)}$ and exercise time T in the Black–Scholes model consisting of the assets B, S.

We are going to obtain expressions for the strategy $(\varphi_B(t), \varphi_S(t), \varphi_D(t))$ separately in the pre-default region $\{t < \tau\}$ and post-default region $\{\tau \le t\}$.

- First, we work in pre-default region $\{t < \tau\}$. Here, (6.14) and (6.15) can be written as

$$\tilde{V}_\varphi(t) = V_\varphi(0) + \int_0^t e^{\Gamma(u)} e^{-ru} \beta_S(u) \sigma S(u) dW_Q(u)$$
$$- e^{-rT} \int_0^t (h(u) - J(u)) \gamma(u) du$$

and

$$\tilde{V}_\varphi(t) = V_\varphi(0) + \int_0^t e^{-ru} \varphi_S(u) \sigma S(u) dW_Q(u)$$
$$+ \int_0^t e^{-ru} \varphi_D(u) D(u, T) \gamma(u) du$$
$$+ \int_0^t e^{-ru} \varphi_D(u) e^{\Gamma(u)} \sigma \alpha_S(u) S(u) dW_Q(u).$$

The right-hand sides will be equal to one another if

$$e^{-rT}(J(t) - h(t)) \gamma(t) = e^{-rt} \varphi_D(t) D(t, T) \gamma(t)$$

and

$$e^{\Gamma(t)} e^{-rt} \beta_S(t) \sigma S(t) = e^{-rt} \varphi_S(t) \sigma S(t) + e^{-rt} \varphi_D(t) e^{\Gamma(t)} \sigma \alpha_S(t) S(t),$$

that is, if

$$
\begin{aligned}
\varphi_D(t) &= \frac{e^{-r(T-t)}(J(t) - h(t))}{D(t, T)} \\
&= \frac{e^{-r(T-t)}(J(t) - h(t))}{e^{\Gamma(t)} e^{-r(T-t)} \mathbb{E}(e^{-\Gamma(T)} | \mathcal{F}_t)} \\
&= \frac{e^{-r(T-t)}(J(t) - h(t))}{e^{\Gamma(t)} V_\alpha(t)},
\end{aligned}
$$

and

$$\varphi_S(t) = e^{\Gamma(t)} \beta_S(t) - \varphi_D(t) e^{\Gamma(t)} \alpha_S(t).$$

To determine $\varphi_B(t)$ observe that

$$V_\varphi(t) = e^{-r(T-t)} H(t) = e^{-r(T-t)} J(t),$$

so

$$V_\varphi(t) = \varphi_B(t)B(t,T) + \varphi_S(t)S(t) + \varphi_D(t)D(t,T)$$

$$= \varphi_B(t)B(t,T) + \varphi_S(t)S(t) + \frac{e^{-r(T-t)}\left(J(t) - h(t)\right)}{D(t,T)}D(t,T)$$

$$= \varphi_B(t)B(t,T) + \varphi_S(t)S(t) + e^{-r(T-t)}J(t) - e^{-r(T-t)}h(t)$$

$$= \varphi_B(t)B(t,T) + \varphi_S(t)S(t) + V_\varphi(t) - e^{-r(T-t)}h(t).$$

It follows that

$$\varphi_B(t) = \frac{e^{-r(T-t)}h(t) - \varphi_S(t)S(t)}{B(t,T)}$$

$$= h(t) - e^{r(T-t)}\varphi_S(t)S(t).$$

- Next, we find the replicating strategy in the post-default region $\{\tau \leq t\}$. Here, formula (6.14) gives

$$\tilde{V}_\varphi(t) = V_\varphi(0) + \int_0^\tau e^{\Gamma(u)}e^{-ru}\beta_S(u)\sigma S(u)dW_Q(u)$$

$$- \int_0^\tau e^{-rT}(h(u) - J(u))\gamma(u)du + e^{-rT}(h(\tau) - J(\tau))$$

$$= \tilde{V}_\varphi(\tau),$$

while (6.15) gives

$$\tilde{V}_\varphi(t) = V_\varphi(0) + \int_0^t e^{-ru}\varphi_S(u)\sigma S(u)dW_Q(u)$$

$$+ \int_0^\tau e^{-ru}\varphi_D(u)D(u,T)\gamma(u)du - e^{-r\tau}\varphi_D(\tau)D(\tau_-,T)$$

$$+ \int_0^\tau e^{-ru}\varphi_D(u)e^{\Gamma(u)}\sigma\alpha_S(u)S(u)dW_Q(u)$$

$$= \tilde{V}_\varphi(\tau) + \int_\tau^t e^{-ru}\varphi_S(u)\sigma S(u)dW_Q(u).$$

The right-hand sides will be equal to one another if

$$\varphi_S(t) = 0.$$

It remains to determine $\varphi_B(t)$. Since

$$V_\varphi(t) = \varphi_B(t)B(t,T) + \varphi_S(t)S(t)$$

and

$$V_\varphi(t) = e^{-r(T-t)}H(t) = e^{-r(T-t)}h(\tau),$$

we obtain

$$\varphi_B(t)B(t,T) + \varphi_S(t)S(t) = e^{-r(T-t)}h(\tau),$$

so

$$\varphi_B(t) = h(\tau) - e^{r(T-t)}\varphi_S(t)S(t).$$

The values of $\varphi_D(t)$ do not matter in the post-default region because $D(t,T) = 0$, and we can take $\varphi_D(t)$ to be given by the same formula as obtained in the pre-default region, that is,

$$\varphi_D(t) = \frac{e^{-r(T-t)}\left(J(t) - h(t)\right)}{e^{\Gamma(t)}V_\alpha(t)}.$$

We have found expressions for $\varphi_B(t), \varphi_S(t), \varphi_D(t)$ in both the pre-default region $\{t < \tau\}$ and the post-default region $\{\tau \le t\}$. This leaves just one difficulty, namely the lack of left-continuity of $\varphi_S(t)$, and hence of $\varphi_B(t)$, at τ. We have already encountered this problem when studying zero-recovery securities in Section 6.3. The remedy applied there also works in the general case. It is to replace $\varphi_S(t)$ by the left limit $\varphi_S(t_-)$ for all $t > 0$. This gives a $(\mathcal{G}_t)_{t\ge 0}$-adapted strategy with left-continuous paths. The resulting expressions for the replicating strategy are collected in the next theorem.

Theorem 6.20
Let h satisfy the assumptions of Theorem 6.18. Then the expressions

$$\varphi_D(t) = \frac{e^{-r(T-t)}\left(J(t) - h(t)\right)}{e^{\Gamma(t)}V_\alpha(t)},$$

$$\varphi_S(t) = \mathbf{1}_{\{t\le\tau\}}e^{\Gamma(t)}\left(\beta_S(t) - \varphi_D(t)\alpha_S(t)\right),$$

$$\varphi_B(t) = h(t\wedge\tau) - e^{r(T-t)}\varphi_S(t)S(t)$$

determine an admissible self-financing strategy $(\varphi_B, \varphi_S, \varphi_D)$ that replicates the contingent claim with payoff $\mathbb{E}(h(\tau)|\mathcal{G}_T)$ and exercise time T.

Proof Almost all the work has already been done. We still need to convince ourselves that the strategy $(\varphi_B, \varphi_S, \varphi_D)$ is admissible. It is because $\tilde{V}_\varphi(t) = e^{-rT}\mathbb{E}(h(\tau)|\mathcal{G}_t)$ is a martingale. Moreover, the above expressions for $\varphi_B(t), \varphi_S(t), \varphi_D(t)$ ensure that $\tilde{V}_\varphi(t)$ satisfies (6.4). By Proposition 6.9, this implies that $(\varphi_B, \varphi_S, \varphi_D)$ is a self-financing strategy. Finally, $V_\varphi(T) = \mathbb{E}(h(\tau)|\mathcal{G}_T)$, hence the strategy replicates the payoff. $\square$

Exercise 6.8 Using the expressions for $h(t)$ and $J(t)$ found in Exercise 6.7, apply Theorem 6.20 to compute an admissible self-financing strategy that replicates the payoff $X\mathbf{1}_{\{T<\tau\}}$ of a zero-recovery derivative security with exercise time $T > 0$. Is this the same strategy as found in Section 6.3?

Exercise 6.9 Find an admissible self-financing strategy $(\varphi_B, \varphi_S, \varphi_D)$ replicating a zero-recovery defaultable bond with maturity S, where $0 < S < T$, and use it to compute the price $D(t, S)$ of such a bond.

Exercise 6.10 Find the price $D(t, S)$ of a zero-recovery defaultable bond with maturity S, where $0 < S < T$, by computing the expectation of the discounted payoff.

6.5 Securities with positive recovery

Our next task is to price a defaultable bond with maturity $T > 0$ that pays compensation in the form of a recovery payment $\eta(\tau)$ at time τ if default occurs before or at maturity.

It will be convenient to consider a derivative security that pays $\eta(\tau)$ at time τ on $\{\tau \le T\}$ and nothing on $\{T < \tau\}$. Combining the price of this security with that of a zero-recovery bond $D(t, T)$ will give the price of the defaultable bond $D_\eta(t, T)$ with positive recovery.

As in Chapter 4, on $\{\tau \le T\}$ the amount $\eta(\tau)$ received at time τ can be invested in the non-defaultable bond until maturity T, when it becomes $e^{r(T-\tau)}\eta(\tau)$. Hence, the payoff of the derivative security is $\mathbf{1}_{\{\tau \le T\}}e^{r(T-\tau)}\eta(\tau)$ at time T.

We assume that η is a non-negative $(\mathcal{F}_t)_{t\ge0}$-adapted process (i.e. related to the market values of default-free securities) with left-continuous paths and such that $\eta(\tau)$ is square integrable under the risk-neutral probability Q, and we put

$$h(t) = \mathbf{1}_{\{t \le T\}}e^{r(T-t)}\eta(t),$$

so the time T payoff of the derivative security can be written as

$$h(\tau) = \mathbf{1}_{\{\tau \le T\}}e^{r(T-\tau)}\eta(\tau).$$

Clearly, h satisfies the assumptions of Theorems 6.18 and 6.20.

Let (β_B, β_S) be an admissible self-financing strategy that replicates the derivative security with exercise time T and payoff

$$m(T) = \mathbb{E}(h(\tau)|\mathcal{F}_T)$$

$$= \mathbb{E}\left(\int_0^\infty h(u)f(u)du \,\Big|\, \mathcal{F}_T\right)$$

$$= \mathbb{E}\left(\int_0^T e^{r(T-u)}\eta(u)f(u)du \,\Big|\, \mathcal{F}_T\right)$$

$$= \int_0^T e^{r(T-u)}\eta(u)f(u)du$$

in the Black–Scholes model consisting of the assets B and S. Then

$$m(t) = \mathbb{E}(h(\tau)|\mathcal{F}_t) = e^{r(T-t)}V_\beta(t)$$

and

$$J(t) = e^{\Gamma(t)}\left(m(t) - \int_0^t h(u)f(u)du\right)$$

$$= e^{\Gamma(t)}\left(e^{r(T-t)}V_\beta(t) - \int_0^t e^{r(T-u)}\eta(u)f(u)du\right)$$

for any $t \in [0, T]$. This leads to the following result.

Theorem 6.21
Let η be a non-negative $(\mathcal{F}_t)_{t\geq 0}$-adapted process with left-continuous paths and such that $\eta(\tau)$ is square integrable under Q. Then the expressions

$$\varphi_D(t) = \frac{V_\beta(t) - \int_0^t e^{r(t-u)}\eta(u)f(u)du - e^{-\Gamma(t)}\eta(t)}{V_\alpha(t)},$$

$$\varphi_S(t) = \mathbf{1}_{\{t\leq\tau\}}e^{\Gamma(t)}\left(\beta_S(t) - \varphi_D(t)\alpha_S(t)\right),$$

$$\varphi_B(t) = h(t\wedge\tau) - e^{r(T-t)}\varphi_S(t)S(t)$$

determine an admissible self-financing strategy which replicates the derivative security that pays $\eta(\tau)$ at time τ on $\{\tau \leq T\}$ and nothing on $\{T < \tau\}$.

Proof By substituting the above expressions for $h(t)$ and $J(t)$ into the formula for $\varphi_D(t)$ in Theorem 6.20, we obtain

$$\varphi_D(t) = \frac{e^{-r(T-t)}\left(J(t) - h(t)\right)}{e^{\Gamma(t)}V_\alpha(t)}$$

$$= \frac{V_\beta(t) - \int_0^t e^{r(t-u)}\eta(u)f(u)du - e^{-\Gamma(t)}\eta(t)}{V_\alpha(t)}.$$

This proves the result. □

Corollary 6.22

The value of the strategy in Theorem 6.21, which replicates the derivative security that pays $\eta(\tau)$ at time τ on $\{\tau \leq T\}$ and nothing on $\{T < \tau\}$, is

$$V_\varphi(t) = e^{r(t-\tau)}\eta(\tau)\mathbf{1}_{\{\tau \leq t\}} + \left(V_\beta(t) - \int_0^t e^{r(t-u)}\eta(u)f(u)du\right)e^{\Gamma(t)}\mathbf{1}_{\{t<\tau\}}$$

for each $t \in [0, T]$.

Proof The value of the strategy $(\varphi_B, \varphi_S, \varphi_D)$ in Theorem 6.21 is

$$
\begin{aligned}
V_\varphi(t) &= \varphi_B(t)B(t, T) + \varphi_S(t)S(t) + \varphi_D(t)D(t, T) \\
&= e^{-r(T-t)}h(t \wedge \tau) - \varphi_S(t)S(t) + \varphi_S(t)S(t) + \varphi_D(t)D(t, T) \\
&= e^{-r(T-t)}\mathbf{1}_{\{t \wedge \tau \leq T\}}e^{r(T-t\wedge\tau)}\eta(t \wedge \tau) \\
&\quad + \frac{V_\beta(t) - \int_0^t e^{r(t-u)}\eta(u)f(u)du - e^{-\Gamma(t)}\eta(t)}{V_\alpha(t)}e^{\Gamma(t)}V_\alpha(t)\mathbf{1}_{\{t<\tau\}} \\
&= e^{r(t-\tau)}\eta(\tau)\mathbf{1}_{\{\tau \leq t\}} + \left(V_\beta(t) - \int_0^t e^{r(t-u)}\eta(u)f(u)du\right)e^{\Gamma(t)}\mathbf{1}_{\{t<\tau\}}
\end{aligned}
$$

for any $t \in [0, T]$. □

Exercise 6.11 Show that the time 0 price $D_\eta(0, T)$ of a defaultable bond that pays 1 at maturity T if no default occurs before or at time T, and otherwise pays recovery $\eta(\tau)$ at the time of default τ, is

$$D_\eta(0, T) = D(0, T) + \mathbb{E}\left(\int_0^T e^{-ru}\eta(u)f(u)du\right).$$

Example 6.23

Let us consider a defaultable bond with positive recovery payment $\eta(\tau)$ at time τ in the event of a default occurring before or at the maturity time T, along with the corresponding **stylised CDS** (stylised Credit Default Swap) composed of:

- the **premium leg** involving continuous payment at a constant rate α from time 0 up to the time of default τ or maturity T, whichever comes earlier;

- the **default leg** paying one unit of cash at maturity T if default occurs before or at time T, reduced by the time T value of the recovery payment $\eta(\tau)$ that a bond holder would have received at the time of default τ.

The default leg payment compensates the bond holder for any loss incurred in the case of default. The rate α, called the **stylised CDS spread**, is set so that the time 0 value of the premium leg is the same as that of the default leg, that is, so that no money is paid when the CDS contract is exchanged at time 0.

We convert the cash flows into payments at time T by investing all payments into the default-free bond. This gives the time T payoff

$$h_{\mathrm{p}}(\tau) = \int_0^{T \wedge \tau} e^{r(T-u)} \alpha \, du = \frac{\alpha e^{rT} \left(1 - e^{-r(T \wedge \tau)}\right)}{r}$$

for the premium leg, and

$$h_{\mathrm{d}}(\tau) = \mathbf{1}_{\{\tau \leq T\}} \left(1 - e^{r(T-\tau)} \eta(\tau)\right)$$

for the default leg.

The following result extends Proposition 4.16, where a stylised CDS on a defaultable bond with zero recovery (rather than with positive recovery) was considered in a simplified market model.

Proposition 6.24
The stylised CDS spread is given by

$$\alpha = r \frac{B(0,T) - D(0,T) - \mathbb{E}\left(\int_0^T e^{-ru} \eta(u) f(u) \, du\right)}{1 - \mathbb{E}\left(\int_0^T e^{-ru} f(u) \, du\right) - D(0,T)}.$$

Proof   The time 0 value of the premium leg is

$$e^{-rT} \mathbb{E}(h_{\mathrm{p}}(\tau)) = e^{-rT} \mathbb{E}\left(\frac{\alpha e^{rT} \left(1 - e^{-r(T \wedge \tau)}\right)}{r}\right)$$

$$= \frac{\alpha}{r}\left(1 - \mathbb{E}\left(e^{-r(T \wedge \tau)}\right)\right)$$

$$= \frac{\alpha}{r}\left(1 - \mathbb{E}\left(e^{-rT}\mathbf{1}_{\{\tau \leq T\}}\right) - \mathbb{E}\left(e^{-rT}\mathbf{1}_{\{T < \tau\}}\right)\right)$$

$$= \frac{\alpha}{r}\left(1 - \mathbb{E}\left(\int_0^T e^{-ru} f(u) \, du\right) - D(0,T)\right).$$

On the other hand, the corresponding value of the default leg is

$$e^{-rT}\mathbb{E}(h_d(\tau)) = e^{-rT}\mathbb{E}\left(1_{\{\tau \le T\}}\left(1 - e^{r(T-\tau)}\eta(\tau)\right)\right)$$

$$= e^{-rT}\mathbb{E}\left(1_{\{\tau \le T\}}\right) - e^{-rT}\mathbb{E}\left(1_{\{\tau \le T\}}e^{r(T-\tau)}\eta(\tau)\right)$$

$$= e^{-rT}\mathbb{E}\left(1 - 1_{\{T < \tau\}}\right) - e^{-rT}\mathbb{E}\left(\int_0^T e^{r(T-u)}\eta(u)f(u)du\right)$$

$$= B(0, T) - D(0, T) - \mathbb{E}\left(\int_0^T e^{-ru}\eta(u)f(u)du\right).$$

If these two values are to be the same, then

$$B(0, T) - D(0, T) - \mathbb{E}\left(\int_0^T e^{-ru}\eta(u)f(u)du\right)$$

$$= \frac{\alpha}{r}\left(1 - \mathbb{E}\left(\int_0^T e^{-ru}f(u)du\right) - D(0, T)\right),$$

which yields the claimed formula for the CDS spread α. $\qquad\square$

We conclude with an example where the structural model is merged with the reduced-form approach.

Example 6.25
Suppose that the value of a company's assets $S(t)$ follows the Black–Scholes dynamics. The company issues debt in the form of a zero-coupon defaultable bond with face value F and positive recovery $R < F$ paid at maturity T. Bankruptcy occurs either at an exogenous time $\tau \le T$ or at the time

$$\theta = \inf\{t \ge 0 : S(t) \le Fe^{-\gamma(T-t)}\} < T$$

when the value process $S(t)$ hits a barrier $Fe^{-\gamma(T-t)}$ with $\gamma > r$, like in the structural barrier model of Section 1.4. The time of default is the earlier of these two times,

$$\eta = \min\{\theta, \tau\},$$

and the bond's payoff is

$$D(T, T) = \begin{cases} F & \text{if } \tau > T \text{ and } \theta > T, \\ R & \text{if } \tau \le T \text{ and } \tau < \theta, \\ Fe^{(r-\gamma)(T-\theta)} & \text{if } \theta \le T \text{ and } \theta \le \tau. \end{cases}$$

The bond price $D(0, T)$ is equal to the expectation under the risk-neutral probability Q of the discounted payoff.

To obtain a specific formula for $D(0, T)$ the joint distribution of τ and θ is needed. For simplicity, we assume that τ has exponential distribution, that is, $F_\tau(t) = 1 - e^{-\lambda t}$ for some $\lambda > 0$, and is independent of the value process S, hence independent of θ. The distribution function of θ was computed in Lemma 1.27,

$$F_\theta(t) = N(d_1(t)) + L^{2\alpha} N(d_2(t)), \tag{6.16}$$

where $\alpha, d_1(t), d_2(t)$ are defined in the lemma, and where

$$L = \frac{Fe^{-\gamma T}}{S(0)} < 1.$$

In this case, the time 0 bond price can be expressed as

$$D(0, T) = e^{-rT} F Q(T < \tau) Q(T < \theta)$$

$$+ Re^{-rT} \int_0^T Q(t < \theta) dF_\tau(t)$$

$$+ Fe^{-\gamma T} \int_0^T e^{-(r-\gamma)t} Q(t < \tau) dF_\theta(t).$$

Substituting the expression $1 - e^{-\lambda t}$ for $F_\tau(t)$ and integrating by parts, we obtain

$$D(0, T) = e^{-rT} F e^{-\lambda T} (1 - F_\theta(T))$$

$$+ Re^{-rT} \int_0^T (1 - F_\theta(t)) \lambda e^{-\lambda t} dt$$

$$+ Fe^{-\gamma T} \int_0^T e^{-(\lambda + r - \gamma)t} dF_\theta(t)$$

$$= e^{-rT} F e^{-\lambda T} (1 - F_\theta(T))$$

$$+ Re^{-rT} \left(1 - e^{-\lambda T} + e^{-\lambda T} F_\theta(T)\right) - Re^{-rT} \int_0^T e^{-\lambda t} dF_\theta(t)$$

$$+ Fe^{-\gamma T} \int_0^T e^{-(\lambda + r - \gamma)t} dF_\theta(t).$$

Finally, by substituting expression (6.16) for $F_\theta(t)$ and applying formula (A.1) in Example A.12 to compute the integrals with respect to $F_\theta(t)$, we can obtain an explicit formula for $D(0, T)$. This is left as an exercise.

Exercise 6.12   Derive an explicit formula for $D(0, T)$ in Example 6.25.

A

Appendix

A.1   Lebesgue–Stieltjes integral

Let $F : [0, \infty) \to \mathbb{R}$ be a right-continuous non-decreasing function.

Definition A.1
For any $b \geq a > 0$ we call $F(b) - F(a)$ the **generalised length** of the interval $(a, b]$.

Example A.2
Here are a few simple examples:
- (i) If $F(t) = t$, then the generalised length of $(a, b]$ is just the length $b - a$ of this interval.
- (ii) If $F(t) = \mathbf{1}_{[x,\infty)}(t)$ for some $x > 0$, then the generalised length of $(a, b]$ is 1 when $x \in (a, b]$ and 0 when $x \notin (a, b]$.
- (iii) If F is the distribution function of a random variable $X > 0$, then the generalised length of $(a, b]$ is $F(b) - F(a) = P(\{a < X \leq b\})$, the probability that X belongs to the interval $(a, b]$.

The following result, which shows that the generalised length can be extended to a measure on Borel sets, can be established by an argument similar to the construction of Lebesgue measure; see [PF].

Theorem A.3

There is a unique measure μ_F on the σ-field $\mathcal{B}((0, \infty))$ of Borel sets on $(0, \infty)$ such that for every $b \geq a > 0$

$$\mu_F((a, b]) = F(b) - F(a).$$

Definition A.4

We call μ_F the **Lebesgue–Stieltjes measure** corresponding to F.

Exercise A.1   Show that for any $a > 0$

$$\mu_F(\{a\}) = F(a) - F(a_-),$$

where $F(a_-) = \lim_{t \nearrow a} F(t)$ is the left-limit of F at a.

The expression on the right-hand side of the preceding formula is the **jump** of F at a, which will be denoted by

$$\Delta F(a) = F(a) - F(a_-).$$

Exercise A.2   Show that for every $b \geq a > 0$

$$\mu_F([a, b]) = F(b) - F(a_-),$$
$$\mu_F([a, b)) = F(b_-) - F(a_-),$$

and, if $b > a$, then

$$\mu_F((a, b)) = F(b_-) - F(a).$$

Definition A.5

The integral $\int_B g d\mu_F$ of a Borel function g with respect to the measure μ_F over a Borel set $B \in \mathcal{B}((0, \infty))$ is called the **Lebesgue–Stieltjes integral** corresponding to F. (For the general definition of an integral with respect to a measure, see [PF].)

Moreover, the Lebesgue–Stieltjes integral can be defined in the case when $F = F_1 - F_2$, where F_1 and F_2 are right-continuous non-decreasing functions. Namely, we put

$$\int_B g d\mu_F = \int_B g d\mu_{F_1} - \int_B g d\mu_{F_2}.$$

Remark A.6

It is well known that $F = F_1 - F_2$ for some right-continuous non-decreasing functions $F_1, F_2 : [0, \infty) \to \mathbb{R}$ if and only if $F : [0, \infty) \to \mathbb{R}$ is a right-continuous function of **finite variation**, which means that the **total variation**

$$V_F(t) = \sup \sum_{i=1}^{n} |F(t_i) - F(t_{i-1})|$$

of F is finite for every $t \geq 0$, where the supremum is taken over all partitions $0 = t_0 < t_1 < \cdots < t_n = t$ of the interval $[0, t]$.

Moreover, the total variation of a function $F = F_1 - F_2$, where F_1 and F_2 are right-continuous and non-decreasing, is given by

$$V_F(t) = V_{F_1}(t) + V_{F_2}(t) = F_1(t) - F_1(0) + F_2(t) - F_2(0),$$

which is also a right-continuous non-decreasing function.

Example A.7

Consider the cases listed in Example A.2:

  (i) $F(t) = t$. Then $\mu_F = m$ is the Lebesgue measure on $(0, \infty)$, and

$$\int_B g \, d\mu_F = \int_B g \, dm$$

    is the Lebesgue integral.

  (ii) $F(t) = \mathbf{1}_{[x,\infty)}(t)$ for some $x > 0$. Then $\mu_F = \delta_x$ is the Dirac measure concentrated at x, and

$$\int_B g \, d\mu_F = \int_B g \, d\delta_x = g(x)\mathbf{1}_B(x).$$

  (iii) F is the distribution function of a random variable $X > 0$. Then $\mu_F = P_X$ is the probability distribution of X, and

$$\int_B g \, d\mu_F = \int_B g \, dP_X = \mathbb{E}(g(X)\mathbf{1}_B(X)).$$

We frequently need to compute Lebesgue–Stieltjes integrals over an interval, and the following notational convention proves convenient.

Definition A.8

For any $b \geq a \geq 0$ the integral of f with respect to μ_F over $(a, b]$ is denoted by

$$\int_a^b f(t)dF(t) = \int_{(a,b]} f d\mu_F$$

and called the **Lebesgue–Stieltjes integral from** a **to** b. Note that the integral is over the interval from a to b open on the left and closed on the right.

The following theorems contain various properties of Lebesgue–Stieltjes integrals used in this book.[1]

Theorem A.9 (integration by parts)
If $F, G : [0, \infty) \to \mathbb{R}$ are right-continuous functions of finite variation, then for each $t > 0$

$$F(t)G(t) = F(0)G(0) + \int_0^t F(s)dG(s) + \int_0^t G(s_-)dF(s)$$

$$= F(0)G(0) + \int_0^t F(s_-)dG(s) + \int_0^t G(s_-)dF(s)$$

$$+ \sum_{s \in (0,t]} \Delta F(s)\Delta G(s).$$

Theorem A.10 (chain rule)
If $g : \mathbb{R} \to \mathbb{R}$ is a continuously differentiable function and $F : [0, \infty) \to \mathbb{R}$ is right-continuous and of finite variation, then the composition $G(t) = g(F(t))$ is right-continuous and of finite variation, and for each $t > 0$

$$G(t) = G(0) + \int_0^t g'(F(s_-))dF(s) + \sum_{s \in (0,t]} (\Delta G(s) - g'(F(s_-))\Delta F(s)).$$

In particular, if F is continuous, then for each $t > 0$

$$G(t) = G(0) + \int_0^t g'(F(s))dF(s).$$

Theorem A.11 (change of measure)
If $F : [0, \infty) \to \mathbb{R}$ is right-continuous, of finite variation, and such that $\int_0^t |g(s)| dV_F(s) < \infty$ and $G(t) = G(0) + \int_0^t g(s)dF(s)$ for all $t > 0$, then G is also right-continuous and of finite variation, and for each $t > 0$

$$\int_0^t h(s)dG(s) = \int_0^t h(s)g(s)dF(s).$$

---

[1] For more information on Lebesgue–Stieltjes integrals and their properties see, for example, D. Revuz and M. Yor, *Continuous Martingales and Brownian Motion*, Springer, 2004.

Finally, using the above properties, we obtain some formulae needed in the proofs of Theorems 1.29 and 1.30, and in Example 6.25.

Example A.12

Let $F(x) = N\left(\frac{-a-bx}{\sqrt{x}}\right)$ for $x > 0$, and $F(x) = 0$ for $x \le 0$, where $a > 0$ and $b \in \mathbb{R}$, and where N is the cumulative standard normal distribution function. Then

$$\int_0^y e^{cx} dF(x) = \frac{d+b}{2d} e^{-a(b-d)} N\left(\frac{-a-dy}{\sqrt{y}}\right) + \frac{d-b}{2d} e^{-a(b+d)} N\left(\frac{-a+dy}{\sqrt{y}}\right),$$

(A.1)

where $d = \sqrt{b^2 - 2c}$, provided that $b^2 > 2c$.

To demonstrate this, we begin by computing the derivative of the right-hand side of (A.1) with respect to y. We have

$$\frac{d}{dy} N\left(\frac{-a-dy}{\sqrt{y}}\right) = N'\left(\frac{-a-dy}{\sqrt{y}}\right) \frac{a-dy}{2\sqrt{y^3}},$$

$$\frac{d}{dy} N\left(\frac{-a+dy}{\sqrt{y}}\right) = N'\left(\frac{-a+dy}{\sqrt{y}}\right) \frac{a+dy}{2\sqrt{y^3}}.$$

Next

$$e^{-a(b-d)} N'\left(\frac{-a-dy}{\sqrt{y}}\right) = \frac{1}{\sqrt{2\pi}} e^{-a(b-d)} e^{-\frac{1}{2}\frac{(a+dy)^2}{y}}$$

$$= e^{cy} \frac{1}{\sqrt{2\pi}} e^{-\frac{1}{2}\left(\frac{a+by}{\sqrt{y}}\right)^2}$$

$$= e^{cy} N'\left(\frac{-a-by}{\sqrt{y}}\right),$$

and similarly

$$e^{-a(b+d)} N'\left(\frac{-a+dy}{\sqrt{y}}\right) = e^{cy} N'\left(\frac{-a-by}{\sqrt{y}}\right).$$

The derivative of the right-hand side of (A.1) is then equal to

$$\frac{d+b}{2d} e^{cy} N'\left(\frac{-a-by}{\sqrt{y}}\right) \frac{a-dy}{2\sqrt{y^3}} + \frac{d-b}{2d} e^{cy} N'\left(\frac{-a-by}{\sqrt{y}}\right) \frac{a+dy}{2\sqrt{y^3}}$$

$$= e^{cy} N'\left(\frac{-a-by}{\sqrt{y}}\right) \left(\frac{d+b}{2d} \frac{a-dy}{2\sqrt{y^3}} + \frac{d-b}{2d} \frac{a+dy}{2\sqrt{y^3}}\right)$$

$$= e^{cy} N'\left(\frac{-a-by}{\sqrt{y}}\right) \frac{a-by}{2\sqrt{y^3}}.$$

This is equal to the derivative of the left-hand side of (A.1) with respect to y. Hence the two sides differ just by a constant. Finally, we need to see that this constant is zero. To this end, we observe that the right-limit as $y \searrow 0$ on the left-hand side of (A.1) is clearly zero. The limit on the right-hand side is also zero because $a > 0$, hence $\frac{-a-dy}{\sqrt{y}}$ and $\frac{-a+dy}{\sqrt{y}}$ tend to $-\infty$, and so $N\left(\frac{-a-dy}{\sqrt{y}}\right)$ and $N\left(\frac{-a+dy}{\sqrt{y}}\right)$ tend to zero.

Exercise A.3 Verify that for any $a \in \mathbb{R}$, $b > 0$, and $y > 0$

$$\int_0^y x dN\left(\frac{\ln x + a}{b}\right) = e^{\frac{1}{2}b^2 - a} N\left(\frac{\ln y + a - b^2}{b}\right),$$

$$\int_0^y x dN\left(\frac{-\ln x + a}{b}\right) = -e^{\frac{1}{2}b^2 + a} N\left(\frac{\ln y - a - b^2}{b}\right).$$

A.2 Passage time and minimum of Wiener process

We denote the minimum of the Wiener process by

$$m^W(t) = \min_{s \in [0,t]} W(s)$$

for all $t \geq 0$. The following result can be found in [BSM].

Proposition A.13
If $c \leq 0$ and $c \leq b$, then

$$P(W(T) \geq b, m^W(T) \geq c) = N\left(\frac{-b}{\sqrt{T}}\right) - N\left(\frac{-b + 2c}{\sqrt{T}}\right).$$

Corollary A.14
A formula for the joint density of $W(T)$ and $m^W(T)$ follows from the above expression by differentiating with respect to b and c:

$$f(b, c) = \begin{cases} \frac{1}{\sqrt{2\pi T}} \frac{d}{dc} e^{-\frac{(b-2c)^2}{2T}} & \text{if } c \leq 0 \wedge b, \\ 0 & \text{otherwise.} \end{cases}$$

Next we consider the process

$$X(t) = vt + \sigma W(t)$$

for some $\sigma > 0$ and $\nu \in \mathbb{R}$, and denote its minimum by

$$m^X(t) = \min_{s \in [0,t]} X(s).$$

Proposition A.13 extends to this case as follows.

Proposition A.15
If $c \le 0$ and $c \le b$, then

$$P(X(T) \ge b, m^X(T) \ge c) = N\left(\frac{-b + \nu T}{\sigma \sqrt{T}}\right) - e^{\frac{2\nu c}{\sigma^2}} N\left(\frac{-b + 2c + \nu T}{\sigma \sqrt{T}}\right).$$

Proof   First we tackle the case when $\sigma = 1$. Then

$$X(t) = \nu t + W(t)$$

is a Wiener process with drift. By Girsanov's theorem (see [BSM]), $X(t)$ is a Wiener process under the probability measure Q with Radon–Nikodym density

$$\frac{dQ}{dP} = e^{-\nu W(T) - \frac{1}{2}\nu^2 T}.$$

We have

$$\frac{dP}{dQ} = e^{\nu W(T) + \frac{1}{2}\nu^2 T} = e^{\nu X(T) - \frac{1}{2}\nu^2 T},$$

hence

$$P(X(T) \ge b, m^X(T) \ge c) = \mathbb{E}_P\left(\mathbf{1}_{\{X(T) \ge b, m^X(T) \ge c\}}\right)$$
$$= \mathbb{E}_Q\left(e^{\nu X(T) - \frac{1}{2}\nu^2 T}\mathbf{1}_{\{X(T) \ge b, m^X(T) \ge c\}}\right).$$

Because $X(t)$ is a Wiener process under Q, Corollary A.14 gives the joint density of $X(T), m^X(T)$ under Q. It follows that

$$P(X(T) \ge b, m^X(T) \ge c) = \int_{-\infty}^{\infty}\left(\int_{-\infty}^{\infty} e^{\nu x - \frac{1}{2}\nu^2 T}\mathbf{1}_{\{x \ge b, y \ge c\}}f(x, y)dy\right)dx$$

$$= e^{-\frac{1}{2}\nu^2 T}\int_b^{\infty}\left(\int_c^{\infty} e^{\nu x}f(x, y)dy\right)dx$$

$$= \frac{1}{\sqrt{2\pi T}}e^{-\frac{1}{2}\nu^2 T}\int_b^{\infty}\left(\int_c^{0 \wedge x} e^{\nu x}\frac{d}{dy}e^{-\frac{(x-2y)^2}{2T}}dy\right)dx$$

$$= \frac{1}{\sqrt{2\pi T}}\int_b^{\infty}\left(\int_c^{0 \wedge x}\frac{d}{dy}\left(e^{2\nu y}e^{-\frac{(x-2y-\nu T)^2}{2T}}\right)dy\right)dx.$$

Consider the case when $c \leq 0$ and $b \geq 0$. Then

$$P(X(T) \geq b, m^X(T) \geq c) = \frac{1}{\sqrt{2\pi T}} \int_b^\infty \left(\int_c^0 \frac{d}{dy} \left(e^{2vy} e^{-\frac{(x-2y-vT)^2}{2T}} \right) dy \right) dx$$

$$= \frac{1}{\sqrt{2\pi T}} \int_b^\infty \left(e^{-\frac{(x-vT)^2}{2T}} - e^{2vc} e^{-\frac{(x-2c-vT)^2}{2T}} \right) dx$$

$$= N\left(\frac{-b+vT}{\sqrt{T}} \right) - e^{2vc} N\left(\frac{-b+2c+vT}{\sqrt{T}} \right).$$

Next, take $c \leq 0$ and $c \leq b < 0$. Then

$$P(X(T) \geq b, m^X(T) \geq c) = \frac{1}{\sqrt{2\pi T}} \int_b^0 \left(\int_c^x \frac{d}{dy} \left(e^{2vy} e^{-\frac{(x-2y-vT)^2}{2T}} \right) dy \right) dx$$

$$+ \frac{1}{\sqrt{2\pi T}} \int_0^\infty \left(\int_c^0 \frac{d}{dy} \left(e^{2vy} e^{-\frac{(x-2y-vT)^2}{2T}} \right) dy \right) dx$$

$$= \frac{1}{\sqrt{2\pi T}} \int_b^0 \left(e^{-\frac{(x-vT)^2}{2T}} - e^{2vc} e^{-\frac{(x-2c-vT)^2}{2T}} \right) dx$$

$$+ \frac{1}{\sqrt{2\pi T}} \int_0^\infty \left(e^{-\frac{(x-vT)^2}{2T}} - e^{2vc} e^{-\frac{(x-2c-vT)^2}{2T}} \right) dx$$

$$= \frac{1}{\sqrt{2\pi T}} \int_b^\infty \left(e^{-\frac{(x-vT)^2}{2T}} - e^{2vc} e^{-\frac{(x-2c-vT)^2}{2T}} \right) dx$$

$$= N\left(\frac{-b+vT}{\sqrt{T}} \right) - e^{2vc} N\left(\frac{-b+2c+vT}{\sqrt{T}} \right).$$

This completes the proof in the case when $\sigma = 1$.

Now, for any $\sigma > 0$,

$$\sigma^{-1}X(t) = \sigma^{-1}vt + W(t)$$

and

$$P(X(T) \geq b, m^X(T) \geq c)$$

$$= P(\sigma^{-1}X(T) \geq \sigma^{-1}b, m^{\sigma^{-1}X}(T) \geq \sigma^{-1}c)$$

$$= N\left(\frac{-\sigma^{-1}b + \sigma^{-1}vT}{\sqrt{T}} \right) - e^{2\sigma^{-1}v\sigma^{-1}c} N\left(\frac{-\sigma^{-1}b + 2\sigma^{-1}c + \sigma^{-1}vT}{\sqrt{T}} \right)$$

$$= N\left(\frac{-b+vT}{\sigma\sqrt{T}} \right) - e^{\frac{2vc}{\sigma^2}} N\left(\frac{-b+2c+vT}{\sigma\sqrt{T}} \right)$$

when $c \leq 0$ and $c \leq b$. $\qquad \square$

As a consequence of Proposition A.15, we can obtain the distribution of $m^X(T)$.

Lemma A.16

For each $c \leq 0$,

$$P(m^X(T) \geq c) = P(X(T) \geq c, m^X(T) \geq c)$$

$$= N\left(\frac{-c + vT}{\sigma\sqrt{T}}\right) - e^{\frac{2vc}{\sigma^2}} N\left(\frac{c + vT}{\sigma\sqrt{T}}\right).$$

Proof   This follows immediately by taking $b = c \leq 0$ in Proposition A.15.
□

We now consider the process

$$Y(t) = y_0 + vt + \sigma W(t) = y_0 + X(t)$$

for all $t \geq 0$, where $y_0 > 0$, and define the passage time

$$\tau = \inf\{t \geq 0 : Y(t) \leq 0\}.$$

The next proposition gives the probability distribution of τ.

Proposition A.17

For each $T > 0$,

$$P(\tau \leq T) = N\left(\frac{-y_0 - vT}{\sqrt{T}}\right) + e^{-\frac{2vy_0}{\sigma^2}} N\left(\frac{-y_0 + vT}{\sqrt{T}}\right).$$

Proof   Since

$$\{\tau \leq T\} = \{Y(t) \leq 0 \text{ for some } t \in [0, T]\}$$
$$= \{X(t) \leq -y_0 \text{ for some } t \in [0, T]\}$$
$$= \{m^X(T) \leq -y_0\},$$

from Lemma A.16 we obtain

$$P(\tau \leq T) = P(m^X(T) \leq -y_0)$$

$$= 1 - N\left(\frac{y_0 + vT}{\sigma\sqrt{T}}\right) + e^{\frac{2vc}{\sigma^2}} N\left(\frac{-y_0 + vT}{\sigma\sqrt{T}}\right)$$

$$= N\left(\frac{-y_0 - vT}{\sigma\sqrt{T}}\right) + e^{\frac{2vc}{\sigma^2}} N\left(\frac{-y_0 + vT}{\sigma\sqrt{T}}\right).$$

□

The following property is needed in the proof of Lemma 1.28.

Proposition A.18

For any $T > 0$ and $y \geq 0$,

$$P(Y(T) \geq y, \tau \geq T) = N\left(\frac{-y + y_0 + vT}{\sigma\sqrt{T}}\right) - e^{\frac{2vy_0}{\sigma^2}} N\left(\frac{-y - y_0 + vT}{\sigma\sqrt{T}}\right).$$

Proof Note that

$$\{Y(T) \geq y, \tau \geq T\} = \{X(T) \geq y - y_0, m^X(T) \geq -y_0\}.$$

By Proposition A.15 with $c = -y_0 \leq 0$ and $b = y - y_0 \geq c$, it follows that

$$P(Y(T) \geq y, \tau \geq T) = P(X(T) \geq y - y_0, m^X(T) \geq -y_0)$$

$$= N\left(\frac{-b + vT}{\sigma \sqrt{T}}\right) - e^{\frac{2vc}{\sigma^2}} N\left(\frac{-b + 2c + vT}{\sigma \sqrt{T}}\right)$$

$$= N\left(\frac{-y + y_0 + vT}{\sigma \sqrt{T}}\right) - e^{-\frac{2vy_0}{\sigma^2}} N\left(\frac{-yy_0 + vT}{\sigma \sqrt{T}}\right).$$

$\square$

A.3 Stochastic calculus with martingales

Here we present the basic tenets of stochastic integration with respect to continuous local martingales.[2] Let $(\Omega, \mathcal{F}, P)$ be a probability space with filtration $(\mathcal{F}_t)_{t \geq 0}$.

Definition A.19
A **Wiener process with respect to** $(\mathcal{F}_t)_{t \geq 0}$ is an $(\mathcal{F}_t)_{t \geq 0}$-adapted stochastic process $W : \Omega \times [0, \infty) \to \mathbb{R}$ such that
   (i) $W(0) = 0$, P-almost surely;
  (ii) for any $0 \leq s < t$ the increment $W(t) - W(s)$ is normally distributed with mean 0 and variance $t - s$;
 (iii) for any $0 \leq s < t$ the increment $W(t) - W(s)$ is a random variable independent of the σ-field $\mathcal{F}_s$;
 (iv) the paths $t \mapsto W(t)$ are continuous, P-almost surely.

Remark A.20
When W is a Wiener process with respect to the filtration $(\mathcal{F}_t^W)_{t \geq 0}$ generated by itself, then we simply say that W is a Wiener process. This is consistent with the terminology in [SCF] and [BSM].

From now on we are going to assume that the filtration satisfies the so-called **usual conditions**, that is, $\mathcal{F}_t = \bigcap_{s > t} \mathcal{F}_s$ and $\mathcal{F}_t$ contains all the sets in $\mathcal{F}$ whose probability P is zero, for each $t \geq 0$.

[2] For more details and proofs of the results see, for example, P. Medvegyev, *Stochastic Integration Theory*, Oxford University Press, 2007, or D. Revuz and M. Yor, *Continuous Martingales and Brownian Motion*, Springer, 2004.

Definition A.21

We say that a process $M : \Omega \times [0, \infty) \to \mathbb{R}$ is a **local martingale** when there is a non-decreasing sequence of stopping times $\tau_n \geq 0$ such that $\tau_n \nearrow \infty$ as $n \to \infty$, P-almost surely, and the stopped process $M_{\tau_n}(t) = M(t \wedge \tau_n)$ is a martingale for each $n = 1, 2, \ldots$.

Proposition A.22

Every non-negative local martingale M is a supermartingale. If, in addition, $\mathbb{E}(M(t))$ is the same for each t, then M is a martingale.

Theorem A.23 (Lévy's characterisation)

Let $M : \Omega \times [0, \infty) \to \mathbb{R}$ be a stochastic process. A necessary and sufficient condition for M to be a Wiener process with respect to $(\mathcal{F}_t)_{t \geq 0}$ is that both M and the process $M(t)^2 - t$ are local $(\mathcal{F}_t)_{t \geq 0}$-martingales with P-almost surely continuous paths and $M(0) = 0$.

Theorem A.24

For every local martingale M with P-almost surely continuous paths, there is a unique process, denoted by $[M, M]$, with P-almost surely continuous non-decreasing paths such that $[M, M](0) = 0$ and $M^2 - [M, M]$ is a local martingale with P-almost surely continuous paths.

Proposition A.25

Let M be a local martingale with P-almost surely continuous paths. Then for every $t \geq 0$ and for every sequence of partitions $0 = t_0^n < t_1^n < \cdots < t_{k_n}^n$ such that

$$\max_{i=1,\ldots,k_n} \left| t_i^n - t_{i-1}^n \right| \to 0 \quad as \ n \to \infty,$$

we have

$$\sup_{s \in [0,t]} \left| \sum_{i=1}^{k_n} \left| M_{s \wedge t_i^n} - M_{s \wedge t_{i-1}^n} \right|^2 - [M, M](s) \right| \xrightarrow{P} 0 \quad as \ n \to \infty,$$

where $\xrightarrow{P}$ denotes convergence in probability.

Definition A.26

We call $[M, M]$ the **quadratic variation** process of M.

Definition A.27

For two local martingales M and N with P-almost surely continuous paths, we also define

$$[M, N] = \frac{1}{4} \left([M + N, M + N]^2 - [M - N, M - N]^2 \right)$$

and call it the **covariation** process of M and N.

Remark A.28

Because $[M, M]$ is a non-decreasing process and $[M, N]$ is the difference of two non-decreasing processes with P-almost surely continuous (hence right-continuous) paths, Lebesgue–Stieltjes integrals with respect to $[M, M]$ and with respect to $[M, N]$ can be defined pathwise, that is, for P-almost every path; see Appendix A.1.

Definition A.29

Let M be a local martingale with P-almost surely continuous paths. By $\mathcal{P}^2(M)$ we denote the family of progressively measurable processes X such that, P-almost surely,

$$\int_0^t |X(s)|^2 \, d[M, M](s) < \infty \quad \text{for each } t \geq 0.$$

Here the integral with respect to $[M, M]$ is a pathwise Lebesgue–Stieltjes integral.

Theorem A.30

Let M be a local martingale with P-almost surely continuous paths, and let X belong to $\mathcal{P}^2(M)$. Then there exists a unique local martingale J with P-almost surely continuous paths such that $J(0) = 0$ and, for every local martingale N with P-almost surely continuous paths,

$$[J, N](t) = \int_0^t X(s) d[M, N](s) \quad \text{for each } t \geq 0,$$

where the integral with respect to $[M, N]$ is understood as a pathwise Lebesgue–Stieltjes integral.

Definition A.31

We call J the **stochastic integral** of X with respect to M, and denote it by

$$\int_0^t X(s) dM(s) = J(t).$$

Remark A.32

The Wiener process W is a local martingale (in fact a martingale) with P-almost surely continuous paths. We have $[W, W](t) = t$, which agrees with the quadratic variation process considered in [SCF]. Moreover, each process X in $\mathcal{P}^2(W)$ belongs to the family $\mathcal{P}^2$ defined in [SCF], and the stochastic integral $\int_0^t X(s) dW(s)$ also agrees with that defined in [SCF].

Remark A.33

If $(\mathcal{G}_t)_{t \geq 0}$ is a larger filtration than $(\mathcal{F}_t)_{t \geq 0}$ on the same probability space $(\Omega, \mathcal{F}, P)$, that is, one satisfying $\mathcal{F}_t \subset \mathcal{G}_t \subset \mathcal{F}$ for each $t \geq 0$, and M has P-almost surely continuous paths and is a local martingale not only with respect to $(\mathcal{F}_t)_{t \geq 0}$, but also with respect to $(\mathcal{G}_t)_{t \geq 0}$, then any process X belonging to the family $\mathcal{P}^2(M)$ of integrable processes with respect to $(\mathcal{F}_t)_{t \geq 0}$ also belongs to the corresponding family of integrable processes with respect to $(\mathcal{G}_t)_{t \geq 0}$, and the stochastic integral $\int_0^t X(s)dM(s)$ is the same under both filtrations.

Theorem A.34

If M is a local martingale with P-almost surely continuous paths, X belongs to $\mathcal{P}^2(M)$,

$$J(t) = J(0) + \int_0^t X(s)dM(s) \quad \text{for each } t \geq 0,$$

and Y belongs to $\mathcal{P}^2(J)$, then XY belongs to $\mathcal{P}^2(M)$ and

$$\int_0^t Y(s)X(s)dM(s) = \int_0^t Y(s)dJ(s) \quad \text{for each } t \geq 0.$$

Remark A.35

We also need to consider processes defined only on an interval $[0, T]$ for some $T > 0$ rather than on $[0, \infty)$. The above definitions and properties readily extend to this case:

(i) We say that M is a local martingale on $[0, T]$ if there is a non-decreasing sequence of stopping times τ_n with values in $[0, T)$ such that $\tau_n \nearrow T$ as $n \to \infty$ and M_{τ_n} is a martingale for each $n = 1, 2, \ldots$.

(ii) For a local martingale on $[0, T]$ with P-almost surely continuous paths we define the family $\mathcal{P}^2(M)$ consisting of progressively measurable processes X such that, P-almost surely,

$$\int_0^t |X(s)|^2 \, d[M, M](s) < \infty \quad \text{for each } t \in [0, T).$$

(iii) For M a local martingale on $[0, T]$ and X belonging to $\mathcal{P}^2(M)$, the stochastic integral $\int_0^{(\cdot)} X(u)dM(u)$ is defined as the local martingale on $[0, T]$ with P-almost surely continuous paths such that for any $t \in [0, T)$ and any $s \in [0, t]$

$$\int_0^s X(u)dM(u) = \int_0^s X_t(u)dM_t(u),$$

where $X_t(u) = X(t \wedge u)$ and $M_t(u) = M(t \wedge u)$ defined for any $u \geq 0$ are the processes X and M stopped at t. Hence M_t is a local martingale with P-almost surely continuous paths, X_t belongs to $\mathcal{P}^2(M_t)$, and the stochastic integral of X_t with respect to M_t can be understood as in Definition A.31.

Select bibliography

The number of books and papers on credit risk is very large, but there are not many presenting a rigorous theory in a mathematical fashion. The main textbook, published a while ago but still remaining the most important source, is this:

T. R. Bielecki and M. Rutkowski, *Credit Risk: Modelling, Valuation and Hedging*. Springer-Verlag, 2002.

The key results from this book are presented in a more accessible manner in the following lecture notes:

T. R. Bielecki, M. Jeanblanc, and M. Rutkowski, *Credit Risk Modeling*. Osaka University CSFI Lecture Notes Series 2. Osaka University Press, 2009.

Among numerous papers, we mention just two related to the approach adopted in the current volume, based on pricing by means of replication:

T. R. Bielecki, M. Jeanblanc, and M. Rutkowski, Hedging of Credit Derivatives in models with totally unexpected default, in J. Akahori *et al.* (eds.), *Stochastic Processes and Applications to Mathematical Finance. Proceedings of the 5th Ritsumeikan International Symposium*. World Scientific Publishing, 2006, pp. 35–100.

C. Blanchet-Scalliet and M. Jeanblanc, Hazard rate for credit risk and hedging defaultable contingent claims, *Finance and Stochastics*, 8 (2004), pp. 145–159.

Out of many less mathematically oriented books on credit risk we select a few, which we believe provide useful supplementary material:

C. Bluhm, L. Overbeck, and C. Wagner, *Introduction to Credit Risk Modeling,* 2nd edn. Chapman & Hall/CRC Financial Mathematics Series. CRC Press, 2010.

D. Cossin and H. Pirotte, *Advanced Credit Risk Analysis*. John Wiley & Sons Ltd, 2001.

G. Löffler and P. N. Posch, *Credit Risk Modeling Using Excel and VBA*. John Wiley & Sons Ltd, 2007.

A. Saunders and L. Allen, *Credit Risk Measurement,* 3rd edn. John Wiley & Sons Ltd, 2010.

P. Schönbucher, *Credit Derivatives Pricing Models: Models Pricing and Implementation.* John Wiley & Sons Ltd, 2003.

Index